D1129914

Highway Economic Series

STUDIES OF HIGHWAY

DEVELOPMENT AND

GEOGRAPHIC CHANGE

WILLIAM L. GARRISON

BRIAN J. L. BERRY

DUANE F. MARBLE

JOHN D. NYSTUEN

RICHARD L. MORRILL

GREENWOOD PRESS, PUBLISHERS
NEW YORK

Foreword

THE HIGHWAY Revenue Act of 1956 recognized that the development of more realistic federal and state policies on highway planning and highway finance required new information on how urban road improvement influences urban communities. Acceleration of investment in highways and the active participation of the Federal Government in user taxation are new developments making new inquiry urgent. The 1956 act directed the Secretary of Commerce to assemble information on user and nonuser benefits from highway improvements. Toward this goal the Bureau of Public Roads engaged in new research, soliciting the services of universities and other agencies.

In response to this solicitation, the University of Washington, through the Washington State Highway Commission, undertook a number of projects. Since 1951 faculty members from many departments and colleges of the unversity have been collaborating in numerous similar studies, all aimed at better understanding of various aspects of the problem of highway transportation in Washington State. The interdisciplinary nature of these researches is a good demonstration of the way in which a state university can best contribute to the solution of problems in its community.

The University of Washington research studies center on the impact of highway facilities on nonusers. They stress the changes in urban arrangement and the extent of uban land utilization that occur in response to highway improvements.

The present volume, *Studies of Highway Development and Geographic Change,* presents findings of investigations of the spatial pattern of shopping centers in their relation to highway improvements, relationships between highway travel and residential and commercial site selection, and the utilization of highway transportation in relation to the arrangement of customer tributary areas and supplying centers at local, regional, and national levels.

A forthcoming volume, *Studies of the Central Business District and Urban Freeway Development,* examines the evolving structure of the central business district and seeks to evaluate its change in relation to burgeoning urban highway networks. Related to and combined with this investigation is a collateral study of the feasibility of charging nonvehicular beneficiaries with responsibility for part of the cost of the principal urban thoroughfares.

The two volumes represent concurrent but autonomous research activities by the individual authors. Each author is, of course, responsible for the research accomplished and reported.

v

Most of the studies were sponsored by an interim highway committee of the legislature or by the State Highway Commission upon the recommendation of the Washington State Council for Highway Research. This state-sponsored activity, which has been described in earlier reports, is intermeshed with the researches described in the present volumes.

R. G. Hennes

University of Washington
Seattle

Preface

DEVELOPMENT of highway facilities is being accelerated considerably today. There is curiosity concerning the implications of this acceleration for the structure of the economy. Knowledge is needed to guide the orientation and capacity of highway facilities as well as public planning of land uses and both public and private investment in the light of accelerated development. In particular, knowledge is needed to guide taxation policies and programs related to highway developments.

Research studies within this volume are designed to provide both general and particular knowledge. Since it is recognized that highway developments are set within the broad context of the geographical structure of our entire social, political, and economic system, the studies stress more than the immediately required information. Fundamental aspects of the geographical structure of our way of life are emphasized, for it is contended that only with this stress can the manifold importance of current highway programs be evaluated. It is essential to keep aspects of emerging patterns of urbanization, resource utilization, land uses, and the like in mind when changes in the network of highways are evaluated. Such emerging patterns are those within which highway-induced changes occur. Hence, in addition to those questions bearing directly on highway networks, questions of geographic structure are fundamental to the subject of highway impact, benefits, or whatever one wishes to term the results of highway improvement.

These studies take the point of view that transportation improvements may be studied both within immediate short-run time periods, when activities and users directly served by that transportation are affected, and also within longer time periods of significant locational readjustment of our entire system such as the widespread transportation-induced suburbanization of cities. In this longer time period a preponderance of the impact of transport improvements falls upon so-called nonusers of the transport facilities. This produces the "nonuser," "indirect" or "nonvehicular" benefits. Demands for information concerning these benefits originally motivated the present research.

The studies in this volume are sample studies of fundamental aspects of the geographical organization of our economic life, the place of highways within this organization, and the influences of highway change. Geographical aspects of the problem are stressed here because they are considered important and have been ignored or treated lightly in much previous work. These aspects are stressed also because although they are complex and difficult they do contain within them most other problems of highway impact and thus are necessary and fruitful areas of investigation. It seems trite to note the space-time continuum within which all action rests.

The studies are grouped into sections which reflect the problem being attacked, the method of approach, and authorship. Chapters 1 and 6 were written by Brian J. L. Berry and William L. Garrison, Chapters 2 and 5 by Garrison, and 3 and 4 by Berry. These six chapters comprise Sections I and II of the volume. Section III is by Duane F. Marble, Section IV by John D. Nystuen, and Section V by Richard L. Morrill. Chapters 3 and 4 are based in large measure upon a previous major inquiry in Spokane, Washington, extended for the present volume. Sections III, IV, and V are parts of continuing larger research inquiries, and the results presented make available their reviews of general information, identification of significant questions, and tentative findings. Continuation of this research will undoubtedly clarify further questions. Chapter 5 is based on re-examination of data from a previous volume by Professor Garrison in collaboration with Professor Marion E. Marts.

Section I is introductory and informational. Chapter 1 introduces the general notion of the place of transportation in the structure of human activities. It orients the reader to highway problems and gives examples of empirical and theoretical research which has contributed to the solution of these problems. This overview permits setting down certain significant questions which are used to summarize the first chapter and to introduce topics of ensuing sections of the volume.

Chapter 2 presents fragmentary notes on the notion of benefits of highway improvements. A definition of benefits is presented which recognizes achievement of efficiencies both within the highway transportation process and within the location structure served by highways as consequences of highway improvements. One portion of the chapter reviews certain standard arguments about benefit questions; another discusses two points of view of the highway development process. The latter illustrates alternate ways of approaching highway problems and how alternate approaches yield alternate notions about benefits. This chapter is a fragmentary treatment of a subject which deserves a larger and more elegant analysis. The information presented is that which seems pertinent to the studies within the volume and to the implications of the volume for Section 210 of the Highway Revenue Act of 1956. The statements are also intended to supplement statements made elsewhere by tax economists and others who have written on benefit questions.

Section II is used to analyze the spatial arrangement of retail business in terms of its association with highway networks. Fundamental questions here are: How are retail business establishments arranged spatially? How is this arrangement related to the transport network? How is it sensitive to transportation changes? The first question, that of the arrangement of business and its relation to the transport network, has previously been studied in a variety of arbitrary ways. Analysis of data for Spokane, Washington, and several other places led to the identification of a fourfold system of business location patterns, which has considerable objective merit in contrast to patterns identified previously. Chapters 3 and 4 examine the status of both theoretical and empirical work available on this topic and present the results of research.

The second question is that of the manner in which this arrangement of business is sensitive to major highway improvements. Empirical data on this topic are very fragmentary, but a recent study of the vicinity of Marysville, Washington, provides one source of information. Chapter 5 presents indices of sensitivity of individual business categories to the major highway reorientation around Marysville. In addition to presenting empirical information, this chapter demonstrates how highway reorientation leads to a variety of changes in travel patterns which, in turn, occasion changes in businesses.

Sensitivity of the spatial structure of business establishments to highway improvements is treated in Chapter 6. Empirical data on individual business categories are examined in the light of the structure of business identified in Chapters 3 and 4; statements regarding change of business structure are made. The question of the influence of zoning practices on business arrangements is treated by examining the case of Spokane, Washington. Influences of the availability of frontage road and intersection sites on the arrangement of business are treated briefly by examining some emerging patterns of land uses at these locations.

The influence of improved highway networks on residential site selection is examined in Section III. Residential site selection takes place within a general structure of both residential and nonresidential land uses, and identification of this general structure is undertaken in Chapter 7. The value of a residential site is a function of many variables, some of which may be classified as location variables. The contribution of location variables to site values is examined in Chapter 8. Three empirical studies are presented which relate location and other data to values. Projections of location trends require fine scale information on the transportation requirements of households of various types, especially knowledge of the stability of travel patterns when available transportation facilities vary. These topics are discussed in Chapter 9, from both theoretical and empirical points of view.

Section IV concerns business establishments as did Section II. But the question of the impact of highway developments on retail businesses is approached using data on travel from residential sites; in this aspect Section IV resembles Section III.

Chapter 10 contains a review of information on movements to obtain goods and services and also reviews information on association of business establishments in response to demands for multiple-shopping opportunities. The analysis of data in Chapter 11 treats both movement to retail stores and location groupings of business establishments. It is stressed that questions of changes in shopping patterns and business associations concomitant with highway network changes are extremely complex, owing chiefly to problems of analyzing multipurpose trips. This analysis is also related to questions of changes discussed in Sections II and III.

Section V treats questions of the arrangements of supplying centers and tributary areas and the sensitivity of these areas to changes in transportation. The section differs somewhat from previous sections by giving specific attention to one activity, medical services, and treating this activity at urban, regional, and national levels. Medical services were chosen for study because of the availability of data. Chapters 12 and 13 treat the questions of the existence of sets and subsets of service-supplying centers and of complementary sets and subsets of surrounding service areas. Empirical data are presented for several areas of study, and general information on the arrangement of trading areas and supplying centers is given. Chapter 14 is an experimental examination of the sensitivity of the system of trading areas and supplying centers to changes in the highway transportation network. The sensitivity problem is to find a new spatial equilibrium under the conditions of decreased transportation time and cost. The analysis permits conclusions regarding changes in service areas and supplying points and benefits to suppliers and consumers of medical services.

Many persons have contributed to this research through discussions with individual authors. In particular, we wish to mention Professor R. G. Hennes and Messrs. David Levin, William C. Pendleton, Jr., and G. P. St. Clair. We also wish to thank Mr. Don Martin of the Outdoor Advertising Association of America and Mr. Victor Pelz of the Traffic Audit Bureau, Inc., for making available the Cedar Rapids data used in Sections III and IV. Mr. William Strange and Miss Elizabeth H. Bailey gave editorial and typing assistance, respectively, and Mr. Merwin W. Parker prepared many of the figures. We hope that the acknowledgment to others who assisted with the research has been made adequately elsewhere. The respective authors take responsibility for their statements and findings.

William L. Garrison

University of Washington
Seattle

Contents

Illustrations

TABLES

FIGURES

Studies of Highway Development and Geographic Change

SECTION I
Introductory

1

Impact of Highway Transportation Innovation

THE DRAMATIC sweep of a modern highway across urban and rural landscapes is becoming a commonplace feature of the American scene. Both prompting and prompted by these modern facilities, the tempo of life is changing; we are traveling further, more frequently, and for more varied purposes than ever before. This is the highway part of the transportation revolution, a revolution with genetic roots in antiquity but one whose effects have snowballed since 1900 and the development of the automobile. To meet these needs we are now planning an investment program of billions of dollars in the interstate highway system and other aspects of federal, state, and local highway programs. Since this transportation revolution is a continuation and acceleration of past trends both in highways and in other modes of transportation, a survey of the information currently available on the impact of transportation expansion in the past can be used to establish bench marks of present knowledge and point out certain needful directions of research.

To motivate this discussion consider the retrospective question: Why have transportation improvements been made? Naturally, transportation improvements have come about because persons have felt that they were worth-while. Improvements have unified large territories into politically cohesive units (e. g., the railroads in the United States in the 1880's and the ancient Roman roads). They have come about to relieve congestion, and then have been made to open up new areas from an economic point of view (railroads in the 1880's in the United States and farm-to-market roads serve as cases in point). Running through all of these various purposes is the obvious feature, specialization. Unifying large territories politically, relieving congestion, opening new economic areas, and other motivating factors in transportation developments were all worth-while because they introduced efficiencies in the form of specialization: armies could specialize, governments specialize, farmers and manufacturers could act as specialized production agents, and routes themselves were specialized as efficient carriers. Individuals or groups of individuals specialized at places to serve larger areas. Certainly, the continued transportation revolution means more specialization. And since this specialization takes place at locations, one key to the study and recognition of significant problems in the study of transportation impact is the observation of transportation-induced changes at these locations. Some of the ways the impact of improved transportation is manifested in distinct places or locations are illustrated below.

The Metropolitan Area Example

The history of the metropolitan area in the United States provides many examples of transport-induced changes. Metropolitan areas found their origins in the emergence of contemporary urbanization marking the transition from subsistence to market economies (Ratcliff, 1953). The transport system became vital to the maintenance of a properly functioning urban organization, since only through the transport system could essential movements of goods and people from one specialized area to another be maintained.

As the transport system has emerged, so have changes occurred in the structure of the metropolitan area. The past half century has seen the transition from horsecar lines and the steam railway to the electric trolley and its applications in elevated and subway lines, and finally to the bus, truck, and passenger automobile. Progressively, distance barriers of time and cost have been lowered. Concomitantly, the metropolitan structure has evolved from the early concentrated form of the metropolis, through the starfish patterns created by populations oriented to the fixed tracks of the earlier transport media, to the more diffuse pattern evolved as the result of centrifugal flows into interstices of the earlier structure because of the flexibility introduced by automobiles and highways (Isard and Whitney, 1949). Around and within major urban centers the highway network has served to realign many community and personal relationships.

Theoretical Discussion

The above description and similar descriptions have been presented many times, yet they remain largely implicit and intuitive. Does any theory exist to make more explicit the role of cheapened transport inputs in effecting structural changes in the system? Isard (1956) has presented a theoretical discussion which facilitates recognition of changes resulting from the cheapening of transport inputs.

He argues that transport inputs are factors of production with positively inclined supply curves and negatively inclined demand curves. Advances in transportation shift supply curves to the right, resulting in lower prices for transport inputs. These may be illustrated with examples to indicate general economic tendencies resulting from lowered prices of transport inputs.

Example of Two Business Establishments

As mentioned above, Isard (1956) has provided an extensive and articulate discussion of the place of transportation in economic activities. From this we borrow and alter an example showing the impact of transportation improvement in a particular and simplified case of two business firms. The example is instructive in that it lays the groundwork for consideration of more complex cases.

Consider two business firms, A and B, who are competing along the line AB where illustrative customers are located at L, M, E, and X as in Figure 1-1. In line with traditional analysis borrowed from economics, the spatial allocation of customers is solved in terms of marginal costs. Take the producer at A. If he serves a customer at the same location, his cost of production is K and he has no transportation costs. Consider the customer at L. The marginal cost of serving this customer is AJ from the standpoint of production and JH from the standpoint of marketing (transportation costs). Customer M is served with still lower production costs, AF, but transportation costs FD for the customer at M are greater than those for the customer at L. In this case the total costs are greater to serve the customer at M than the customer at L. It is important to recognize that these are variable costs associated with the firm and with transportation to the customer.

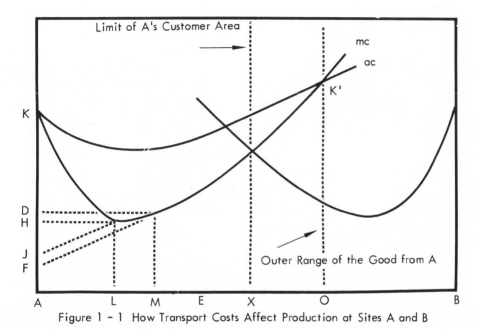

Figure 1 - 1 How Transport Costs Affect Production at Sites A and B

With this explanatory discussion in mind, pertinent features of this figure are quite transparent. The outer range of A's possible production is the point O (where the average cost curve [ac] crosses the marginal cost curve [mc]), given average cost pricing practices by A. This is a point that has been termed the outer range of a good (Berry and Garrison, 1958c). To serve the area beyond O another producer is warranted. The location of the competitive producer at B is such that the actual marketing territory of A extends only from A to X, and beyond X, B controls the market.

Another Example

Consider A and B in the case where transportation improvements are made within their vicinity. In Figure 1-2, A and B are redrawn and B's production costs are slightly greater than A's (it is easier to assume unequal production costs than equal production costs). Part B of the figure shows how A and B's market areas are mapped out in two-dimensional (geographical) space. A decrease in transportation cost occurs and shifts in producing and consuming areas also occur. It is extremely interesting to note and not intuitively obvious that the whole situation in production and in the size of the consuming areas changes. Output has changed at producing sites and areas of consumption also change.

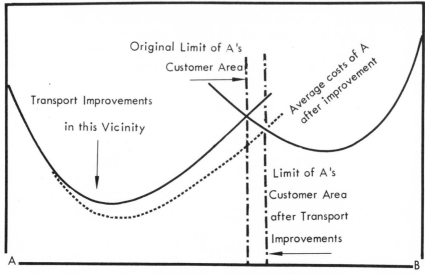

Cross-Section View -- How an improvement in transportation affects
(Part A)   A's and B's production.

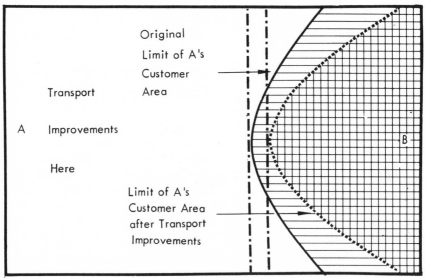

Figure 1 - 2 Vertical View -- Rearrangement of A's and B's customer areas
(Part B)   owing to transport improvements.

The Benefits

    Only in a very special case can improvements in transportation be of equal advantage to all areas. The very nature of specialization and the geographic system means a differentiation of benefits among places, and therefore it is very important to keep in mind the fact that benefits from transportation improvements are almost always differentiated geographically as well as in more obvious ways. By more obvious ways we refer to differentiation among industries, among kinds of highway users, among kinds of transportation, and the other things. A key feature of research reported in the ensuing chapters of the present study is the emphasis given to the geographic differentiation of benefits from transportation improvements. The argument is that industry benefits, as well as other benefits, are fairly well known, in a general way. What is needed to apply this general knowledge to specific cases is knowledge of the geographic differentiation of benefits.

More General Cases

The nature of the impact of transport change has been discussed from the point of view of two producers in a simple market area. This "Robinson Crusoe-like" example is useful for motivating interest in geographic aspects of transportation problems and in identifying questions to be answered. It is unsatisfactory in that it is not clear how the larger system (one with more centers, more industries, and more tributary areas) would function or how problems identified by a larger system might be different from those in a simple example.

A chief omission of the two-producer example is that it does not allow for location shifts within the industry. Another is that the technique used in discussing the two-producer cases did not allow consideration of alternate pricing policies. Just what happens in a many-producer case is by no means completely understood.

A second example may elaborate the latter remark. The case in point is that of wholesale grocery centers and trading areas. The nation is divided into many of these and there are many firms in the industry, presumably more than one in each wholesale center. The example contains many areas and many producers (service centers) and many industries (wholesale grocery establishments). The pattern displayed by these areas is one that has evolved over a long period of time. With transportation improvements competing centers have varied in their ability to offer service to tributary areas, and the fortunes of individual firms as well as centers have waxed and waned with these changes. Also, the number of centers and complementary areas is an outcome of the transportation available. When transportation was poorly developed, there were many small centers (admittedly, the industry was much more primitive than it is today). Concomitant development of transportation and the wholesale grocery industry has resulted in the areas and centers shown on the map today. Transportation improvements will continue. Consequently, concomitant developments will continue in the wholesale grocery industry. Will these developments mean new maps of wholesale grocery centers and tributary areas? In particular, will it mean fewer but larger centers of wholesale grocery industry and larger complementary trading areas?

Intuitively, and in terms of our historical experience, the analyst may answer this question with a tentative yes, although he would do well to point out that these changes will likely be by no means absolute. New types of wholesaling will undoubtedly develop to take advantage of less expensive transportation, and much of what is present today will remain as a residual pattern imbedded within the pattern resulting from larger developments.

Other illustrations of the general impact of transportation developments and the scale and substitution effects resulting from transportation improvements are provided by consideration of land uses and consumer movements. In the case of the impact of a transportation improvement on land uses, a substitution effect will extend the spatial extent of the system of land uses. Cheapened transport inputs are substituted for other inputs, especially those at inferior sites with respect to the improved transportation system. Scattered ubiquitous patterns of land use are thereby transformed into increasingly concentrated ones and the interdependence between areas through the transport system increases. Progressively, differentiation and selection between sites with superior and inferior resources and locations lead to greater geographical specialization of land uses. Also, inputs in general at favored sites will be substituted for inputs in general at inferior sites. Accompanying scale effects upon levels of consumption are too well known to require further discussion.

Cheapening of transport inputs likewise results in diminished costs of consumer movement. Consumers are able to maintain given levels of social contact and consume more, for example, enjoying such additional benefits as suburban residence. Scale effects of this nature in part account for the phenomenon of dispersion of urban population and settlement to peripheries of metropolitan areas in the past half century. Decentralization is also partly due to substitution effects since transport inputs are substituted for other inputs, as is evidenced by increasing proportions of consumer budgets devoted to transportation.

Summary

The brief discussion just completed has identified the impact of current changes in transportation as merely the continuation of specialization processes of long standing. The discussion has also given a fragmentary view of how the specialization process works from the standpoint of business firms, urban land uses, interurban relations, and consumers.

A more concrete view of the influence of transportation improvements is needed for appreciation of the significance of the types of the research reported in this volume. The remaining portion of the chapter is used to review some empirical work which serves as substantive background for the research undertakings.

Trends Related to Transport Improvements

A great deal of empirical work is available to verify the impact of transportation improvement. Particular attention has been given to urban and suburban change concomitant with improvements, but broad works are available on the role of transportation improvements in national and international developments, and works are available on such restricted cases as small towns and rural communities. Attention has also been directed to the important topics of highway finance, engineering problems of bridge construction, and safety, as well as numerous other topics.

While a review of all works related to transportation improvements would be pertinent in this introduction to research on highway improvement problems, it is hardly practicable. The compromise of arbitrarily selecting a few works for review is adopted and they are used as a device to point up some apparent results for verification and, if verified, for more careful codification of transport changes.

Change Relating to Metropolitan Areas

An analysis by Isard and Whitney (1949) suggests that changes in metropolitan areas resulting from major transport innovations include increasing specialization of land uses and concentration of economic life with attendant increases in interdependence between economic activities and land uses, with congestion in central areas, and with decentralization to the peripheries of expanding metropolitan centers.

Do these hypothesized changes have empirical validity? The question must be answered in light of available evidence. Of course, empirical studies are many and diverse, and consequently the comparisons presented here are intended to be suggestive rather than comprehensive. Only a sample of the many studies available is considered.

R. D. McKenzie (1933) observed increasing concentration of the population of the United States during the first half of the automobile period, and he presented a detailed discussion of the significance of urban aggregation. The increasing focal role of the large city is indicated in Table 1-I.

TABLE 1-I
CENTRALIZATION OF THE POPULATION

Year	U.S. Population (Millions)	Counties with 25 Per Cent of the Population	Counties with 50 Per Cent of the Population	Counties with 75 Per Cent of the Population
1910	92	39	312	1,068
1920	106	33	250	992
1930	123	27	189	862

McKenzie associated this concentration with the development of large focal regional aggregates called "metropolitan communities." He argued that the major contributing causes of the growth of these foci were:

1. the increasing role of secondary and tertiary activities in a specialized society--activities which are better performed at the center rather than extensively in the field;
2. the shrinking of space by modern communications, with "center" activities performed for wider and wider areas as the rural-urban dichotomy declined in importance. He held that transport improvements have been responsible for the welding of previously self-sufficient areas into single functioning economic units, and that the innovation of the automobile has brought previously independent towns and villages into functioning metropolitan units.

Isard and Whitney (1949) have specified some of the economic effects of this process of development of metropolitan units. They show that a rearrangement and differentiation of functions among urban sites have taken place during the automobile era.

Of course, the establishment by the automobile of a site selection effect with respect to consumer functions has already been anticipated theoretically. Increasing population mobility necessitates a repatterning of retail service institutions. At the same time the consumer hinterland around the metropolitan center is extended, the resistance to interurban movement is lessened by reduction of the time-cost dimensions of local travel. The metropolitan center as the confluence of transport flows tends to usurp increasingly those functions which are performed efficiently only for wide service areas supplying either heavy volumes of business or sufficient potential customers to support specialized products.

Empirical findings of the study are that, for 100 cities of similar retail structure in 1890, the automobile has evoked a differentiation of retail functions between central and outlying cities, viz.:

1. core cities have appropriated the retail trade in general merchandise and apparel categories at the expense of cities within a 20-mile radius;
2. sales of immediate consumer goods, such as food, furniture, automobiles, lumber and building materials, drugs, and jewelry, show least tendency to be usurped;
3. total per capita sales figures indicate a general retail site selection in favor of metropolitan centers over cities of 10,000 population within a 20-mile zone of metropolitan center.

Rural Areas

In studies of the rural areas of the Prairie Provinces of Canada C. C. Zimmerman (1938) noted the concentration of facilities in, and growth of, larger trading centers with improvements of transport facilities which enabled them to reach over much wider areas than previously. A progressive division of labor between smaller and larger areas was observed to have been facilitated by diminutions in the time-cost dimensions of distance in the system. Also, smaller centers not previously located on trade routes shifted to more favorable transport facilities. Zimmerman discussed at length the increased complexity of social organization and the increased interdependence and more complex trading relations of larger and smaller centers associated with improvements in transport facilities and an improving standard of living.

The rise of the larger rural community with more complete facilities was likewise seen in two studies of Walworth County, Wisconsin, by Galpin (1915) and Kolb and Polson (1933). They observed that as better transport facilities were provided it became possible to benefit from increased economies of scale because people were able to travel further to obtain their needs. Local facilities were by-passed for larger scale facilities with greater breadth and depth. Specialization and growth of areal interdependence through the highway net were seen to be integral parts of the development of the county, facilitated by the innovation of the automobile.

James S. Burch (1957), in an interesting study of the secondary road problem in North Carolina, emphasized the many changes that have come about in rural areas concomitant with highway improvement. Burch lists 16 indices of change including such varied topics as yields of agricultural products, number of voters in rural areas, and degree of rural industrialization. All of these indicators suggest an improved quality of life in rural areas, and Burch makes the point that these improvements were made possible by improvements in secondary roads.

Thompson and Madden (1953) have provided a study of relations between socioeconomic characteristics, place of employment, residential location, and the quality of highway service in a rural area. In particular, their study points up how location of residences is influenced by the quality of the road system and amount of travel. They also link the amount of travel to family income and socioeconomic status. In general, the higher the socioeconomic status the greater the demand for highway services. But these authors stress the complex relationships between type of employment, family characteristics, and other variables which bear on highway use in rural areas.

These studies in rural areas indicate the existence of a complex set of interurban relations which result in the changing status of rural communities as the highway system is improved. Also, they illustrate the complex relations that exist between communities and their surrounding rural areas. For a review of empirical research on highway impact in rural areas, the reader is referred to Garrison's study (1956) which stresses problems of measuring the value of highway improvements to rural areas.

The Rural-Urban Fringe

The decentralization of metropolitan areas into the rural-urban frings is documented dramatically by the consideration of the growth of metropolitan areas in the United States. A map of the city of Chicago, for example, attests to the sprawl of that metropolitan community into surrounding areas. The spokelike extension of the city is aligned with transportation facilities. Thus it is easy to see the impact of routes on the rural-urban fringe. In recent years the highway and automobile have been chief among the transportation influences aligning the arrangement of metropolitan centers.

Many authors have stressed the many problems attendant upon this sprawl of metropolitan areas into the rural-urban fringe. In regard to processes, Martin (1953) wrote:

> Under the impact of the private automobile, the pattern of small dependent communities strung along interurban lines has given way to a mass residential decentralization independent of the railroads, out to areas of settlement which were virtually non-existent prior to the great "explosion" of cities which accompanied the development of rapid transportation and mass communication systems and related urban conveniences beyond the city's political boundaries. The continued expansion of the population concerned in this peripheral, or fringe area, is one of the most significant of population trends.

The accelerated suburbanization of American life identified by Dessel (1957) verifies a continuation of these trends.

Particular attention to the role of the highway in suburbanization was the feature of studies by Wheeler (1956) and Moore and Barlow (1955). Wheeler's study was from the point of view of the working of the land market along the rural-urban fringe and the relation of this land market to modern highway facilities. In particular, Wheeler pointed out how property values in the suburban fringe tended to vary with the availability of modern transportation facilities. Moore and Barlow were concerned with an area in the vicinity of East Lansing, Michigan, and stressed the difficulties residents in rural areas encounter as a result of the suburbanization movement, but advantages were also indicated. The analysis was set within a matrix of the availability of transportation to the rural area, and it demonstrated how the impact of metropolitan development is conditioned by the availability of transportation.

Small Area Studies

One approach to identification of changes contingent upon changes in the highway system has been the study in smaller areas of particular highway improvements and their nearby impact. In comparison with the studies reviewed earlier, these studies are less general in the sense that they refer to small areas, particular kinds of highway improvements, and particular results. On the other hand, they are motivated by interest in highway improvements and not by more general considerations. The studies thus yield rather pointed results. This is unlike studies discussed earlier.

The study of Camarillo, California, by investigators of the California Division of Highways and reported in the volume by Garrison and Marts (1958a), serves as an example of small area studies and information available from these studies. In this case a by-pass highway was constructed paralleling and close to the road it replaced (Figure 1-3). The new highway was opened in March, 1954. The California investigators compared the levels of business sales and property values during the year before the facility opened with business sales and property values after the new facility was opened. Special attention was given to cafes and bars, and service stations. The conclusion the investigators reached was that commercial investments have continued to be made in the community at a satisfactory rate and the value of property fronting the old highway "increased at a higher rate than similar property elsewhere in the community."

Most of the studies in the volume by Garrison and Marts mentioned above are of the character of the Camarillo study, but several more general studies are also summarized in the volume. An example of such a study is that of the central expressway and associated land values in Dallas, made by the Texas Transportation Institute (Adkins, 1957). This study was more general in the sense that the facility was larger and the zones of influence studies extended some distance away from the facility itself. It shares the character of the small area studies, however, in being restricted to a particular type of improvement and impact. This study clearly shows the zonal impact of highway developments.

In a second study Garrison and Marts (1958b) present the influence of a particular highway improvement in terms of many indicators. The subject improvement was a by-pass highway around Marysville, Washington. The study showed how the levels of land values for six types of land uses varied in relation to their location with respect to the new facility and how the sales of 30 types of business tended to vary with respect to the location of establishments. This study indicates the great variability of impact depending on the type and location of the activity being affected.

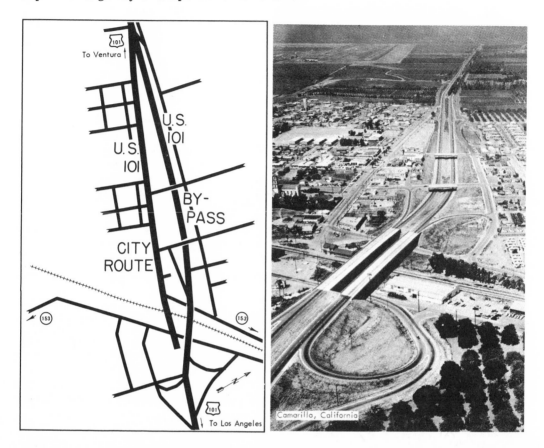

Figure 1-3. In Camarillo the new highway facility was constructed parallel and close to the old facility (photograph from the California Division of Highways).

Dynamics of Trends

It may not be assumed that a particular type of transportation improvement will lead to a particular set of effects. There is no assurance, for example, that widening a two-lane highway to a four-lane facility will result in traffic flows of some given amount, land value increases of some given amount, population changes of some stated amount, or some other given result. All of these results depend on whether or not the activities served by the transportation are conducive to more, or a different, use of transportation.

There are several ways controls over the dynamics of trends made possible by highway improvement may be recognized. One may think of the *demand* for transportation by households, firms, and land uses, or by other classifications of activity. Knowledge of demand is needed to answer questions of the sort: if more transportation is available how much will be used? Also, problems of pricing the transportation service depend on demand considerations. Stated as a substitution problem by Isard (1956), the question is: how much will transportation costs be substituted for some other costs of the activity? The analysis problem is to quantify production and consumption characteristics of particular activities with respect to transport inputs.

A complicating feature of demand analysis, so far as the highway system is concerned, is the great increase in demand occurring because of the operation of the economy at higher and higher levels. One aspect of this growth has been the explosive increase in population. In the case of households and the transportation they will require in the years ahead, it is necessary to have knowledge of the unsatisfied demand of households, and how this unsatisfied demand will change in the future. Good examples of considerations such as these are available from highway-planning and city-planning documents.

Still another complicating feature of transportation analysis is the complex system of spatial interrelationships of which the transportation system is an integral part. Characteristics of spatial systems are available from Friedmann (1955), who emphasized how the spatial organization of activities have changed along with the economic development of the Tennessee Valley. He lists some main tendencies:

1. growing importance of central locations--increasing specialization in the Tennessee area with increasing concentration and centralization and improved accessibility to central location;
2. rapid growth of city-regions resulting in a more tightly knit locational system replacing the local self-sufficient pattern which previously existed in the area;
3. excessive concentration balanced by the greater flexibility of the new transport system allowing decentralization into metropolitan areas of urban activities and residences;
4. differential patterns of growth and welfare in the system with expansion at the center and lag at the periphery, being less accessible and less central. Once established, these differential patterns of growth and welfare tend to perpetuate themselves.

Knowledge of structural change on the order of that which Friedmann has provided is certainly an essential requirement for the analysis of transportation improvements. The transportation system clearly plays a role in these changes and perhaps a critical role due to the competitive character of many activities. The demand for transportation is present and a tendency toward a change in spatial structure is present. The timing and magnitude of transportation developments tend to determine the directions in which these latent tendencies for change are directed. What this is saying is that the nature of the transportation system determines the type of impact that it will have, but that it is not sufficient for change to have a change in the transportation system. It is necessary that the demand for transportation be present and this, in turn, is conditioned by a complex system of interrelationships.

Study Topics

Recognition that knowledge of demand for transportation and the spatial structure within which this demand operates is essential to the interpretation of transportation developments conditioned the research studies undertaken and reported in the ensuing portions of this volume. Impressions about the character of the impact of transportation improvements gained from empirical and theoretical materials such as those reviewed earlier also helped determine the research studies undertaken and the strategy used. In brief, the strategy in the research was that of verifying whether or not the presumed impact takes place, and then attempting to codify the impact, providing it could be verified in the first place. At the same time emphasis was given to identifying the technical conditions under which the verified and codified impact takes place.

Certain of the notions regarding the character of transportation-induced changes that oriented the research are listed below. They serve as a summary of this chapter and as an introduction to the subjects of ensuing chapters.

1. Knowledge of the structure of transportation use by households, business firms, and other activities using the transportation system is central to all questions about transportation impact. A great deal of work has been done on this subject and more work is included within the present volume. Stress here is on relations between the use of transportation and urban land uses and their spatial arrangement. The guiding notion is that transportation conditions the arrangement of urban uses and a change in availability of transportation changes the arrangement of land uses.

2. It is recognized that the desire for increased transportation services varies for each and every person and for each and every group of persons or land uses, and that the way one activity uses transportation may influence how another activity uses transportation. The notion of variations and especially the existence of interdependence among variations also influenced the study greatly. In particular, attention was given to retail land uses and activities and how they vary in sensitivity to transportation changes, alone, as integrated groups, and as related to transportation uses by households.

3. Many workers have stressed the importance of the notion that transportation services are set within a complex hierarchical and spatial structure of metropolitan areas, cities, villages, towns, and so on. Orientation and capability of transportation influence the character of this structure. This notion also guided research reported here. The existence of structure is given attention in the analysis of the transportation relations of retail business and also in a study of the provision of medical services at different places from various orders of centers. An attempt is made to project complex interrelationships between the system of city areas, the demand for transportation by the activity of medical services, and the utilization of transportation by households in the procuring of medical services.

There are endless other facets to the transport revolution and no claim is made that notions identified above exhaust the field. They are singled out for study because they are interesting and because their further interpretation should go far toward illuminating the nature of transportation-induced changes.

2

Notes on Benefits of Highway Improvements

THE STUDIES within this volume are motivated by the notion that their subjects are extremely significant ones. The studies are set within and contribute to a broad body of information on the spatial structure of human activities. The arrangement of land uses, urban centers, transportation routes and processes of movement, specialization, and production that accompany these arrangements are examples of the content of the broad field of information. On a practical level there are implications for urban and regional planning, marketing, and other activities. The materials are particularly applicable to the development and evaluation of highway systems. There is a direct relation between this research and the Highway Revenue Act of 1956 (especially Section 210), in which Congress asked for a study of the benefits of the accelerated highway construction program of the Federal Aid Highway Act of 1956. Congress' interest in benefits is from the point of view of taxation but it is obvious that the study of benefits is related to the study of the demand for transportation and is of interest for the planning of the highway system as a part of a larger integrated transportation system.

This chapter is used to present some fragmentary notes on benefits in the light of the direct relation of these studies to the problems presented by the Highway Revenue Act of 1956.

The discussion to follow is definitional, argumentative, and expository. Just what is meant by these terms is explained shortly. The discussion should also be characterized as rather elementary. Questions of benefits align themselves with broad questions of man's whole affairs. It is completely obvious that we do not act perfectly, owing perhaps to our motivation and to the elementary state of our information and understanding. The elementary treatment of benefits in part mirrors this improper motivation as well as a lack of understanding and information.

The first part of the discussion defines benefits by the locations where they occur and uses the dichotomy of vehicular versus nonvehicular benefits. A confusing aspect is that benefits may be transferred from one place to another and this problem of transferred benefits is included to complement the definition of benefits by location. It is recognized that benefits may be defined using a variety of strategies. The method used to recognize benefits (and develop the highway system) determines where benefits occur, how large they are, and how well they may be transferred. Here benefits of highway improvements are defined in a particular way. This definition does not end all definitional problems, but benefit questions are clarified when it is recognized that benefits realized will vary depending on benefits recognized and used as a guide for action.

There is a great deal of controversy regarding what benefits are and what public policy regarding benefits should be. This controversy is treated in portions of the ensuing text which are argumentative. The author has heard and participated in many public discussions which began with the elaboration of a rather obvious point: in public and private activities benefits that flow from that action should guide planning the size, place, and timing of the undertaking. The particular case here is the highway system, but these same points have been made regarding other kinds of activities (e.g., water resource development).

While the general argument in the preceding paragraph is readily accepted, it is also easily attacked by making rather standard points. The author has left meetings of the discussion of benefits with the impression that participants at the close of the discussion had only a fragmentary recognition of benefit problems and their importance. This was true because of the skillful counterarguments that can be used to destroy simple benefit concepts without constructively supplying substitutes.

The strategy in the present discussion is to recognize these counterarguments (those arguments against using benefit information as the basis for action) and argue as to their substance. These arguments are elaborated in the second portion, the argumentative portion, of this chapter.

The remainder of the chapter attempts an examination of the highway system as a whole. It elaborates how the allocation of capacity to portions of the system and the resultant traffic assignments and benefits are intimately bound together. Two cases are discussed. The first is a case in which the highway planner treats capacity allocation strictly from a point of view of the movement of traffic and traffic alone. The second case is one in which broader questions concerning the utilization of resources are used to guide highway investment decisions.

The expository discussion illustrates the nature of the several types of benefits that flow from investment in the highway system, and the discussion illustrates how benefits that flow from investments relate to the method used in allocating capacity within the highway system. Previous discussions chiefly have been limited to a single intersection or a single link in a transportation system. It is presumed that what is said about a single intersection or a single link in the transportation system can be generalized to the whole system. From the published literature, just how this is to be done is by no means clear. The latter portion of the present discussion does attempt a generalization to the whole highway network.

The definitional, argumentative, and expository discussions follow.

Identification of Benefits

It is moot why the literature is less than voluminous on the subject of highway benefits. Policy decisions by legislators and highway officials appear to have been made adequately without reference to really fine scale information on benefits from highway improvement programs. Perhaps there has been widespread endorsement of the proposition that continued investment in the highway system has been warranted by its underdevelopment. This may have led to the impression that benefits warrant practically any investment or development in any magnitude or direction. Perhaps it has been the recognition of general public good stemming from transportation improvements (and the willingness of the general public to pay for highway improvement, especially through the gas tax) that has allowed the existence of benefits to be assumed rather than defined and identified quantitatively.

Some rather serious technical problems confront the researcher who concerns himself with benefits, so perhaps this difficulty of dealing with problems has contributed to lack of work with the topic. Just how difficult analysis is becomes clear when one realizes the interdependence of segments of the highway network. It is not enough simply to take the cost of an improvement (say, adding an over-pass at a congested intersection) and compare this cost with vehicular benefits derived (say, saving time, fuel, and vehicle wear and tear). An improvement in one place in the network may divert traffic from other places and otherwise make flow through the whole network more expedient. It is necessary to examine benefits through the whole network in order to make judgments about a particular improvement. Anything less than an examination of the whole network introduces errors into the analysis. It is not surprising that benefit analysis has presented the analyst with almost intractable problems.

But intractability of problems does not mean that they should be ignored. Large sums of money are being invested in highway capacity and highway equipment. This investment is leading to more efficient operation of the economy and, indeed, the reshaping of the economy from a geographical point of view. The strategic position of places tends to change along with concurrent changes in the ease of getting from one place to the other. These changes are highway impact, benefits, fulfillment of demands, or whatever one wishes to call them. They are obviously extremely important.

Vehicular and Nonvehicular Benefits

It is common to recognize vehicular and nonvehicular benefits from highway investment. Vehicular benefits are recognized as those which occur while using the highway and nonvehicular benefits are those occurring elsewhere. By elsewhere one refers to the lands fronting on the highway as well as those separated from the immediate zone of the highway, and to users of these properties. It is clear that both kinds of benefits exist. Any road improvement increases the ease of movement and benefits the highway user. An example may be cited as proof of the existence of nonvehicular benefits. It is well known that a highway improvement may improve drainage patterns; this results in improved drainage of properties in the vicinity of the highway. Here is a nonvehicular benefit from highway improvement.

It might be mentioned that effects of highway improvements may be negative or positive. There is certainly no assurance that effects are always positive. In the case of drainage changes, such as those just cited, negative effects might well be the case.

Benefits: The Geographical Framework

The existence of benefits of vehicular and nonvehicular categories is not open to question. Questions that need to be answered theoretically and quantitatively are questions of the magnitude and exact geographical position of these benefits and related costs. Questions of magnitude and location are especially difficult to treat because vehicular benefits may be transferred to locations where nonvehicular benefits occur.

A miniature example will serve to illustrate the phenomenon of transferred benefits. A road is improved from the city into a suburban area and costs of driving from residential sites in the suburban area to the city and back are markedly reduced. These are vehicular benefits and result from demand for better transportation. In real cases, benefits are not generally captured by tolls on the facility set at rates which exactly capture the savings of each user. Rather, the lower cost of vehicular transit is reflected in inflated property values in the area served by the highway. To the extent that benefits that could be captured by vehicular tolls are not so captured, vehicular benefits appear at nonvehicular locations, or rather, at locations which imply the nonvehicular character of benefits.

It is somewhat more complex to think of the transfer of vehicular benefits received by a commercial carrier but not collected by tolls. In this case, the vehicular benefits are transferred to producers and consumers of commodities carried and are evidenced in the market area organization of services, sales, and industries and in the competitive arrangement of zones within which customers travel for the purchase of goods.

Vehicular, Nonvehicular, and Reorganization Benefits

The above paragraphs have tended to define vehicular benefits as any benefits which could be collected by a toll. (It was recognized that vehicular benefits generally are not so collected and appear elsewhere in the economy.) Consideration of just what this toll might be needs to be made before the nature of vehicular benefits and other benefits may be viewed clearly. This turns itself into a problem of social objectives vis-à-vis the highway system and problems of pricing systems with which economists tussle. Again, an informal and extremely simple example will illustrate some of the difficulties. The reader who is impatient with simple examples may skip directly to Table 2-1 where the benefit statement is summarized.

Suppose a highway improvement is undertaken, the yearly cost of which is one million dollars. Yearly savings to present vehicular users are also one million dollars, so the project is justifiable (leaving aside questions of alternative investment). The vehicular benefit is one million dollars and a toll is set exactly to capture that benefit. There is no increase in number of trips and none of the vehicular benefits is passed on to nonvehicular locations.

But other methods of setting tolls are available. Suppose the designer of the toll has available the information that if the charge to vehicular users is reduced by some amount per trip, say thirty cents on the average, the number of trips per year would increase, say double to two million, and this would maximize the income from tolls. This is not an unlikely bit of intelligence in our society where lower transportation costs have meant larger firms, lowered cost of production, and more goods transported over longer distances. The operator of a toll road would be quick to utilize this information. It is clear that under these conditions benefits of a nonvehicular type would be generated at the places which originate and receive commodities being carried over the highway and that these nonvehicular benefits would very depending on tolls charged. The per trip cost of transportation is down, which results in more goods being shipped, more efficient production, and so forth. This is saying that the increased use of the lower cost transportation results from possibilities for the more efficient operation of vehicles, firms, and households, given better highway facilities. In the instance just elaborated there are *vehicular, nonvehicular,* and *reorganization benefits.* Reorganization benefits occur because nonvehicular benefits require reorganization of activities using the facility. One should also recognize that investment costs are associated with each type of benefit.

Total Benefit

It is now possible to define what is meant by the benefit of a highway improvement. The benefit of a highway improvement *is not* taken to be that amount arrived at when vehicles are charged a toll exactly equal to their yearly savings, for this is the special case when nonvehicular benefits are zero. Also, it *is not* taken to be that sum that maximizes income from tolls. The benefit of an improvement is taken to be a combination of vehicular and nonvehicular benefits. Maximum benefit is that combination which yields a maximum return from the highway facility.

The benefit situation in regard to the example highway improvement is summarized in Table 2-I. Before the improvement was undertaken, some sort of facility existed connecting the places being served. Vehicular and nonvehicular benefits at levels x and y, respectively, existed, and there also existed a spatial organi-

TABLE 2-I
BENEFIT STRUCTURE OF THE EXAMPLE ROAD IMPROVEMENT

Situation	Vehicular Benefits	Reorganization Benefits	Nonvehicular Benefits	Total Vehicular and Nonvehicular Benefits of Facility
Before improvement	x	yes	y	x + y
Vehicle tolls capture all time, etc., savings of the improvement	$1,000,000 + x	None additional	y + 0	$1,000,000 + x + y
Toll income maximized	$1,400,000 + x	Spatial reorganization	$y + y_1$	$1,400,000 + x + y + y_1$
Joint maximization of vehicular and nonvehicular benefits*	$x_2 + x$	Additional spatial reorganization	$y + y_2$	$x + y + x_2 + y_2 =$ maximum

*See the discussion in the text.

zation of activities utilizing the facility. In the first instance the facility is improved and a toll is applied which exactly equals time, cost, and convenience values of the improvement. There are no benefits other than those captured by the toll, and the benefits associated with the facility are the benefits of the original facility plus the toll.

In the next case, a toll is concocted to maximize income from the improvement.

The per vehicle toll is somewhat lower than that which captures all benefits of time, cost, and convenience and there is an increase in traffic. The increase in traffic is a result of the reorganization of activities using the facility and increased levels of output in the amount y_1.

But maximization of income from a toll would be desired only in the limited case of the operation of a toll facility. It ignores the structure of benefits of the original facility (perhaps an alternate route), nonvehicular and reorganization benefits are only incidental, and costs are not considered. What is desired is to find quantities x_2 and y_2 which form a joint maximum over all costs (cost of highway improvement, reorganization cost, etc.) subject to technical, budget, or other restrictions.

This is the benefit target to be striven for in the design of facilities and in the design of accompanying taxation.

Earlier the notion of transferred benefits was introduced. It may be seen from the table that the notion of transferred benefits is a complex one. Benefits may be thought of as being transferred from the nonvehicular category to the vehicular category (for example, in the case where income from tolls was maximized) as well as in the other direction (for example, if there were no tolls).

One is tempted to try to make statements about the relative magnitudes of x_2 and y_2 and associated costs. One difficulty here is that the discussion used the convenient fiction of the average. Another is that almost nothing is known at the empirical level regarding volumes of traffic and tolls charged for different highway uses. It may be that for many, if not most, users tolls in the form of gasoline taxes and other taxes on nontoll highways are much smaller than those which could be charged to maximize the combined or total benefit from highway improvement. However, lack of empirical reference leaves quantitative aspects of the discussion somewhat up in the air so far as defining total benefits as something that can be recognized from relationship between tolls charged and traffic generated in connection with nonvehicular benefits.

Related Statements

It should be stressed that many authorities claim that the proper definition of benefits is in terms of the highway user alone and not in the broad terms of the paragraphs above. Brownlee and Heller (1956) have remarked that "The alleged benefits of highways to those who do not use them directly are primarily illusions arising from failure to charge highway users appropriately for the services provided by the highway system." Brownlee and Heller also cite others who take the same point of view. This author hesitates to attempt to interpret the words of others, yet the quoted statement is so at variance with the previous discussion that some explanation seems essential. Perhaps the crux of the difference is in the view of demand for highway services. The present author assumes that a decrease in the charge for a highway service will greatly multiply the amount of highway use concomitant with spatial reorganization and higher levels of production and consumption. Brownlee and Heller apparently view demand for improved facilities as stemming chiefly from the transportation activity, rather than from both the transportation activity and activities at places served by that transportation.

It is interesting that Beckmann *et al.* (1956) describe a hypothetical demand curve which is concave downward (see Chapter 9 of the present volume). Decreases in average trip cost increase the number of trips per unit of time at a decreasing rate. Beckmann's careful discussion differs from the present discussion, however, by excluding shifts in the location structure served by the highway facility. In the present discussion shifts in location structure are taken to be immediate and marked.

Many authors have referred to problems of benefit definition and analysis using notions of primary and secondary benefits. A recent volume by McKean (1958) presents these notions and also contains an excellent summary of problems of benefit analysis. In addition to primary and secondary benefits, McKean identifies a set of spillover effects. Spillover effects refer to external economies or diseconomies which are of two types: technological and pecuniary.

It would be necessary to lengthen this discussion extensively to identify properly these notions of primary and secondary benefits and spillover effects and to couch the discussion of highway benefits in these terms. Although not possible here, such an extension would be very desirable for it would bring concepts and experiences from resource development to bear on highway problems.

It may be stated here, however, that the consequences of highway development identified in the previous discussion may be thought of either as direct benefits and spillover effects or as direct benefits. One way to make this point is to view the highway service (the road and carriers using the road) as if it were a single firm. A highway improvement is of direct benefit to this firm. Also, the improvement occasions certain efficiencies within organizations shipping or moving via the highway service (manufacturing, households, etc.). These latter efficiencies are external to the highway service and are spillover effects. McKean notes that it is quite appropriate to consider these spillover effects (and costs) in evaluating the value of investment. All benefits would be direct benefits if the activities dichotomized in the statements just made were considered a single activity.

McKean's volume contains a number of comments on the meaning of economic efficiency and the use of market prices. The reader is referred to that volume for an elaborated discussion of appropriate criteria for investment decisions.

By way of summary, some of Isard's (1956) ideas regarding the results from cheapening transport inputs are listed below:

1. More transportation inputs consumed. These were treated in the previous discussion as vehicular benefits which may be captured by toll.
2. Substitution of superior for inferior sites and resources. These are the spatial reorganization benefits identified in the earlier discussion.
3. Greater consumption of other inputs in the system, apart from the greater use of transportation inputs. This operation of activities served by the highway at higher levels was identified previously as nonvehicular benefits.

The identification of these benefits in combination is a major challenge to the researcher concerned with highway matters and with matters of the geographical organization of political, social, and economic facets of society.

Notes on Some Arguments Against Benefit Considerations

Many have had extreme misgivings concerning use of benefits as a guide for action in developing highway systems. In the following, statements which seem to be the basis of these misgivings are evaluated. It is the position here that statements against benefit considerations are generally spurious, and this is revealed in the strategy of arguing against these statements. For simplicity, benefits are discussed without direct reference to associated costs.

It should be noted that this discussion by no means results in firm proof that consideration of benefits and action on the basis of benefits is warranted. The discussion rejects arguments which are made to prove exceptions to rules that benefits exist and may be measured, that benefit considerations result in equitable and optimum policy, and so on. Rejection of these particular arguments does not prove that there is nowhere an exception to one or more of the benefit ideas.

The Double Counting Argument

In discussions with both lay and technical groups, the author has often encountered the argument that attention to nonvehicular benefits is, at best, specious double counting of benefits. It is argued that nonvehicular benefits (observed, for example, as increases in property values) should not be brought into benefit calculations along with materials on travel time savings, lower cost of vehicular operation, and other items related to vehicles. It is observed that nonvehicular benefits are actually anticipated savings in time, vehicle operating cost, etc., and these savings are already counted when vehicular benefits are computed. This observation certainly contains elements of truth. It is the assertion from this that is faulty--that no attention should be given to nonvehicular benefits.

The element of truth in the statement above is a result of faulty estimations. When properly estimated, as presumed in Table 2-I, counting of vehicular and nonvehicular benefits does not lead to double counting.

It is true that improper counting of benefits of vehicular and nonvehicular types *may* lead to double counting and thus an overstatement of highway benefits. It is false to say that counting benefits of nonvehicular and vehicular types necessarily leads to double counting.

The difference between the two previous statements may well be that between an adequate and efficient highway system developed in terms of its total benefits to the economy and an incorrect system developed in terms of a partial evaluation or an overevaluation of benefits.

The Absurd Benefit Argument

Consider the following example: It is well known that improved transportation facilities in the form of roads together with the level of living enjoyed in the United States have led to an efficient automobile manufacturing industry. It is reasonable to presume that continued development of the highway system will lead to larger markets and an even more efficient automobile manufacturing industry. The competitive position of the automobile manufacturer is bettered and, among other effects, a higher level of living is enjoyed by persons employed in the industry.

Should this higher level of living be considered as a benefit when highway investments are programed? Very often the position is taken that such considerations are ridiculous. Support for this is the argument that the high level of living of workers in automobile plants is a joint effect of innovations in the industry, efficient unionization, and endless other things, as well as improvements in highway transportation. Who is to say how much of the credit road improvement is to get? Put as a demand question, who is to say how much of the demand for a higher level of living is to be met by development of the highway system?

The essence of the argument is that because a benefit is hard to measure it should not be measured. Stating a complex sequence of benefits simply emphasizes how difficult benefit measurement is. The complex sequence has no bearing on the question of whether or not benefits exist.

To follow the line of reasoning that only obvious benefits are significant may result in overlooking significant motivation for highway investment. It is presumed that it is a national objective to develop aspects of the economy which lead to the greatest total benefits. Knowledge of the net effect on the whole economy is extremely pertinent if we are to have highway improvements toward the objective stated.

The No Significance for Taxation Argument

Another argument in support of the proposition that nonvehicular benefits should not be evaluated is the argument that these benefits could not possibly have consequences for taxation policy. One statement in support of this is that the benefits lead in so many directions that taxation schemes would be so complicated as to be impracticable. Another statement looks at the more obvious of nonvehicular benefits, such as increments in land values in areas served by new highway facilities, and claims that it is politically inexpedient to devise a taxation scheme to capture these benefits even roughly. The latter is a claim not supported by empirical experience (see Hennes, 1956) and a claim somewhat weakened by the fact that many urban streets and rural roads have been improved with funds from property owners: in the case of urban streets by direct assessments, in the case of rural roads by millage on property.

The first statement that benefit structure is so complicated as to defy concoction of a tax schedule is another case of a reasonable observation leading to a false conclusion. It is traditional that, when benefits of activities are so widely spread as to defy collection by particular earmarked taxes, the benefits are recognized in general taxation. If the statement is true, the action called for is not behaving as if there are no benefits from the activity. The benefit may be recognized in general taxation.

The No Need To Do It Argument

Another argument against considering nonvehicular benefits is the argument that their consideration is entirely unnecessary to an effective highway program. The argument begins by observing that the American people have expressed their desires as to the kind of highway system they want (expressed in the Highway Act of 1956). Calculations indicate that this highway system may be purchased out of user taxes, and thus there is no need to consider nonvehicular benefits. There are several places where one goes astray here.

The assertion that the American people have expressed their desire as regards the highway system is open to some interpretation. The statement is not criticized on the basis that each person has not been asked what he desires. This is too much to expect. The assertion is criticized on the basis that knowledge of benefits of all types is essential to stating what type of highway system is desired. Really informed statements regarding the desired size of the highway system could not possibly be made because no knowledge of a reasonable sort is available on the benefit structure of highway programs.

What is available for planning purposes are statements of desires produced by informed state highway officials on the basis of extremely limited information regarding benefits. As noted, these are not statements of desires made in the full knowledge of the benefit structure of highway improvements. Knowledge of benefits is essential to a statement of the problem.

A word could be said about the notion that vehicular taxes are large enough to pay for highway improvements. Even if the highway system finally identified as that desired is unchanged from that presently estimated, it does not follow that, because income from vehicular taxes is large enough to pay for the facility, this is the procedure that should be followed. Relationships between tolls charged to collect vehicular benefits, nonvehicular benefits, and geographical changes were discussed earlier. Perhaps some toll structure should be adopted other than that at present in use (fuel taxes, etc.). Perhaps tolls should be lowered to cause greater efficiencies at places served by the highway system and greater joint benefits. Perhaps charges to the highway user should be raised. This is a question that is completely open, given the present state of knowledge. But it is a pressing one since, as pointed out by Brownlee and Heller (1956), the Federal Government is invading the user tax field.

The Double Standard Argument

It is sometimes noticed that businesses in the private sector of the economy arrange their activities and charge prices only in terms of the immediate user of the product of the activity. The question is raised: should activities in the public sector, such as highway development, be allowed to utilize the benefits in a manner denied activities in the private sector? In particular, the argument is sometimes strengthened by reference to the problem of competition between highway transport and rail transport (other kinds of transport could also be brought into the argument) and to the manner in which the double standard might affect prices and an efficient allocation of traffic between various methods of transportation.

There are several ways to talk against this point. One would be to question the degree that it is an empirical fact that activities in the private sector are concerned only with the immediate user of the service of the activity. One can cite a counterexample in the matter of railroads and the development of the transportation network in the western United States. Lands were acquired, town sites were laid out and, in general, anticipated and real nontoll benefits played a tremendous role in the development of that industry.

On the other hand, there are many cases where activities in the private sector are developed in terms of the immediate user of the activity. A utility developing a hydropower site, for example, would develop the site at the level justified by benefits in the form of electricity that may be sold to consumers. Benefits in the form of flood control, recreational lakes, and the like, may be created quite incidentally to the objective of the operation of the utility. The utility is unable to capitalize on these values and for this reason is not warranted in taking them into account in the construction of facilities.

The actual point is that many activities in the private sector of the economy are not organized as effectively as they might be. Perhaps matters should be so arranged that the above utility could take flood control and recreation into account. In defense of the private sector, one must admit that in many cases lack of effective organization is a matter over which the operators have little or no control. Railroads, for example, have been required by law to cease the operation of coal mines. Here is a case where the value of a site is at least in part a result of the investment in railroad trackage, yet the railroad is unable to take advantage of this benefit. This type of restriction is unfortunate in many respects, but it is hard to see how it has any bearing on highway development programs.

The Pay for Negative Effects Argument

It is sometimes pointed out that in addition to benefits of a nonvehicular sort there are negative results from highway improvements. When a town is by-passed, for example, cafes often do less business than prior to the construction of the by-passing highway. One can imagine the argument: "If we are to use the inflation of land values at the intersection to pay for the highway, then we should pay the land owner for the deflation of land values at cafe sites in by-passed towns. This type of payment is not traditional in our country. Thus we must discard the nonvehicular benefit point of view."

Such a statement as that in the preceding paragraph misses important aspects of the nature of benefits. One aspect missed is the association of geographical reorganization with highway improvements (see the discussion of benefits earlier in the chapter). Reorganization itself is not a benefit. The benefit is the higher level of output of the activity subsequent to reorganization.

The statement assumes, for another thing, that consideration of nonvehicular benefits is proper only if a taxation program is practicable in exact correspondence to benefitees. There are reasons other than possibility of taxation in exact correspondence to benefits for the consideration of nonvehicular benefits.

Not to consider nonvehicular benefits would cause all development consideration to depend on vehicular benefits which when maximized may not meet the objectives of maximum total benefits identified in the earlier discussions of benefits. Innovation of a method to recognize the total character of benefits from highway improvements is needed. Lack of such innovation is no basis for an out-of-hand rejection of broad benefit concepts.

The argument from the point of view of taxation is only part of the argument, at any rate. Total benefits should be considered for purposes of making proper decisions about highway systems. Whether or not difficult aspects of negative effects, reorganization, etc., are cleared up at a taxation level does not destroy this point.

Relationships Between Highway Development Procedures and Benefits

This portion of the analysis is chiefly to illustrate the intimate relationship between the procedure used to develop the highway system and the impact from that development. Two cases are used. The first is termed the highway planner's case because it illustrates in a crude way the approach to planning the highway system from the point of view of serving traffic most efficiently. The second case is termed the regional planner's case because it illustrates an approach to planning the highway system in terms of efficient regional resource development. As emphasized before, the approaches are incompletely stated and do not do justice to the sophisticated character of planning. But they serve the purpose of illustration which is desired here.

For convenience the symbols utilized in the analysis are defined in the section below. These will be used in the form defined for the analysis of the highway planner's case and will be altered somewhat for the analysis of the regional planner's case. The symbols are presented and are followed by an elaboration of the two cases. Conclusions from the two cases do not appear until both have been elaborated.

Features of a Highway Network

Definitions of traffic flows, capacity, and capacity cost are presented in a list below:

x_i It is presumed that traffic may be measured for the ith route in an amount x_i. The measurement might be the per day flow of tons of goods, passengers, or passenger cars. The routes are those recognized in the highway network for planning, traffic control, and investment purposes, i = 1 to n, which are the n routes in the system.

k_i The highway development problem is to select capacity increments to be allocated to different places in the highway system during the development. Capacity inputs are made on the ith route, in the amount k_i.

There is great variation in definitions of capacity.
Some authors speak of capacity using supply functions (". . . the supply of trip opportunities at various costs . . .," Beckmann *et al.*, 1956) and others use other definitions. A definition is used here which is in terms of the ability of a route to carry its characteristic traffic. Capacity required to handle traffic is measured by associating with each route a constant y_i which indicates the capacity of that route in terms of its characteristic traffic per unit of time. $y_i = 1$ for an even traffic stream (at optimum speed) and is less than 1 when traffic varies from time to time.
Traffic is in full use of the capacity of the ith route when the equation

$$k_i = \frac{x_i}{y_i}$$

holds.

p_i Associated with each possible improvement is the cost of that input identified by p_i, the dollar cost of a capacity increment on the ith route. Total capacity costs are

$$\sum_i p_i k_i$$

A word is appropriate regarding the layout of the highway network subject to analysis. The network consists of (1) links between intersections and (2) intersections. Every place that desires and uses a transportation connection is connected with every other place in the system by either a direct or an indirect route. Viewed grossly, one might treat cities as the intersections on the transportation network and connections between cities as links. Capacity is to be added to the links in the network.

The reader may have noticed that over-the-road transport costs were not defined for links on the transportation system. Variable transportation costs are excluded in the interest of simplification of the analysis.

The Highway Planner's Case

The highway planner uses knowledge of present traffic and traffic projections and of the capacity of existing facilities, and technical knowledge of engineering feasibility and cost to project and design highway systems. His goal is to meet demands of projected traffic most efficiently. This means that he balances cost of construction, cost of travel time, and congestion cost to keep joint cost at a minimum. He uses devices of traffic assignment, estimates of the value of travel time, and the like in the design of efficient facilities. These complex and not completely codified procedures are reproduced in a simplified way in the discussion below.

Capacity is to be installed where it will be demanded by projected traffic. The amount of traffic on a route at the beginning of the study period is adjusted upward to that to be accommodated by the end of the improvement period. This amount is recognized here as

x_i* This is an estimate of traffic in the direction of the ith route and is used as a starting place for the analysis.

The problem of adding capacity is not merely that of supplying capacity for the traffic in the places where traffic is forecast. This is because flows on the transportation network are interdependent and they may be diverted from route to route. At the most elementary level this traffic diversion must be recognized. At more advanced levels one might recognize increased cost of over-the-road transportation due to traffic diversion, but this is not included in the present simple model. For a simple example of diversion consider a network consisting of two links between a single origin and a single destination. An apparent deficit of capacity on both links might best be remedied by installing capacity on just one of the routes. The installed capacity on the improved route would be great enough to handle the excess traffic on that route as well as that diverted from the other route.

In order to handle this feature of traffic diversion, it is necessary to recognize the existence of traffic diversion constants. Capacity additions on routes other than the ith route would divert traffic to and from the ith route in the quantity

$$\sum_{j \neq i} a_{ij} k_j$$

which is the total diverted traffic due to capacity increases of k_j.

Allocation of capacity on links must recognize these diversion constants (a_{ij}), which are plus when traffic is added and minus when traffic is diverted, as well as the projected traffic on the routes (x_i*). Associated with the projected traffic is a projected amount of capacity

$$k_i* = \frac{x_i*}{y_i}$$

The rule for allocating capacity is to find k_i's which solve the set of equations.

$$y_i k_i = y_i k_i* + \sum_{\substack{j \\ j \neq i}} a_{ij} k_j \qquad i = 1 \text{ to } n$$

These equations require that the projected traffic on each route be accommodated either by capacity on that route or by capacity on other routes to which it is diverted.

There is the possibility of timing capacity increases in a manner which would meet the target at the end of the development period but which would create intolerable amounts of traffic diversion during the period of development. It might be intolerable, for example, to begin the first year's work on the Interstate Highway System by tearing up one-fifteenth of all interstate routes and diverting all of this traffic to local roads and secondary state highways. Capacity increments must be programed in a way to keep the capacity deficits that occur during the construction period below some tolerable limits. This leads to an "intolerability" rule to be followed during the allocation of capacity to routes. This rule creates interesting problems but adds little to the question of who benefits from the improvements, so it will not be discussed here. It is mentioned above as an example of dynamic problems that appear in actual situations but which are not discussed here.

The engineer programing a sequence of improvements must have cost notions in mind. For one thing, the objective of bringing capacity in balance with traffic demands is to be met at the least possible cost. That is, of the many ways that might be available to bring capacity into line with demands, the method is to be chosen that brings this about at a minimum cost. The objective recognized is to minimize

$$\sum_i p_i k_i$$

This is a simplification of the planning engineer's actual objective which likely also would include minimizing over-the-road transportation cost. This simplification of the problem has been mentioned before.

In summary, the problem developed is to minimize the total cost of capacity subject to the rule stated earlier that all traffic will be accommodated either by a direct installation of capacity or diversion by capacity installation on another route. Put another way, that vector of k's is desired which minimizes the objective function subject to the restraint equations.

The Regional Planner's Case

The regional planner has a somewhat different view from that of the highway planner. He is chiefly concerned with the best use of the region, so his statements regarding highways are in terms of regional resource use. He proceeds by making projections of desired resource use and traffic associated with resource use and uses this information together with information on highway cost to project transportation routes where they are feasible. The simplified discussion below reproduces, in a crude way, decision problems that face the planner of a regional highway system.

Subscript notation will be altered from that in the preceding problem. In the present problem, the amount x_{ij} refers to the flow of traffic along the route from place i to j. It is understood that the "from i to j" meaning is attached to subscripts of other variables and constants that appear in the problem. For example, P_{ij} refers to the cost of a unit of capacity on the route from place i to place j. The convention of interpreting capacity in terms of traffic using a quantity y_i adopted in the previous problem will also be used in the present problem. y_{ij} interprets pecularities of the traffic between i and j from the point of view of capacity. The capacity is exactly equal to traffic when the condition

$$k_{ij} = \frac{x_{ij}}{y_{ij}}$$

holds.

Concern is with allocating capacity to routes in light of the fact that capacity allocated will change the flow characteristics of routes. The flow equilibrium equations relate to flow caused by production and consumption at places i and j. Concern is with equations in the form

$$\sum_j x_{ij} = \sum_{\substack{j,kl \\ kl \neq ij}} b_{ij,kl} k_{kl} \qquad i = 1 \text{ to } n$$

where $b_{ij,kl}$ is a constant to be defined shortly. The equation specifies how a capacity increase from k to l will change the flow of traffic from the ith place to all other places. The information in mind is that places on the transportation network act competitively. Any change in the competitive position of one place (the kth place has improved capacity on the route to the lth place) will have some effect on the other centers (the ith place will change its flow to the jth place). The coefficient $b_{ij,kl}$ interprets the amount of effect. This is a constant defining the effect on the competitive position for production in the ith place of an improvement in transportation from the kth place to the lth place. It indicates how the production pattern tends to change when the highway system is changed.

Traffic changes related to consumption may be displayed by equations similar to those used in discussing the production process,

$$\sum_j x_{ji} = \sum_{\substack{i,kl \\ kl \neq ji}} c_{ji,kl} k_{kl} \qquad i = 1 \text{ to } n$$

The coefficient $c_{ji,kl}$ interprets the manner a capacity change from k to l will affect place i's imports.

The present problem is in terms of route capacity choices in relation to resource utilization. The planner has production possibilities at each place in mind and has also made decisions regarding levels of consumption at each place. Maximum production at place i is a known constant, B_i, in units of outflow traffic associated with that production (e. g., truck loads of coal). Desired consumption at place j, C_j, is also a known constant and is in traffic flow units. Using these amounts, the planner's rules for traffic assignment and capacity allocation are

$$B_i \geqslant \sum_{kl \neq ij} jkl^{b_{ij}}, kl^{k_{kl}} \qquad i = 1 \text{ to } n$$

$$C_j \leqslant \sum_{kl \neq ij} i, kl^{c_{ij}}, kl^{k_{kl}} \qquad j = 1 \text{ to } n$$

These equations require that nowhere will capacity be added to cause production above the maximum possible or consumption below the minimum allowed. There are also other restrictions which would apply in an actual case. However, the equation system identified above is adequate for the present analysis.

A simple objective for the planner would be to minimize

$$\sum_{ij} p_{ij} k_{ij}$$

which together with the equations stated above pose the problem.

Evaluation

The highway planner's case and the regional planner's case just presented are each oversimplified, and the two represent a rather artificial division of points of view regarding highway development. The chief omission in the oversimplification is failure to include cost of transportation along with capacity cost as cost to be minimized. The simple statement of capacity is also an oversimplification.

The two cases represent an artificial division in the sense that both resource use and traffic accommodation should be included within a unified decision scheme for highway development. Those concerned with highways might claim that both considerations are actually included within the planning process.

Yet in practice there are certainly many cases where the opposing points of view are not merged. The author knows of one case where a problem in bridge location was studied both by an urban planning organization and a highway planning organization, and different solutions were reached as to the most desirable location of the facility. Each solution was quite legitimate, given the point of view of the research agency and the circumstances under which it was obligated to work. The regional planning agency's basic information consisted of desired projections of land use at each terminus of the bridge. The highway planner's information was in terms of present traffic and its projection in terms of present trends. Certainly neither agency was wrong. What is needed is a merger of the points of view of both agencies.

More important to the present analysis are variations in benefits that arise from the two approaches. The planners' cases could be stated and treated as linear programing problems. It is well known that associated with minimization programing problems are maximization problems in which the matrix of constants of the restraining equations of the minimization problem appear in transposed order as the matrix of constants of the restraint equations of the maximization problem. These maximization problems use a new set of variables not previously defined in the discussion.

The associated maximization problems will not be developed formally here in order to avoid making the study unduly complex and since the problems are over-simplified and their computation is not envisioned. For sample applications of problems of this type to transportation problems the reader is referred to Garrison and Marble (1958) and to Beckmann *et al.* (1956). As regards the formal construc-tion of the problems it is important to note that if an optimum solution exists for the problem of allocating capacity to networks (the minimization problem) then an optimum solution exists for the maximization problem. It is also important to note that the variables of the related maximization problems are in monetary terms and may be thought of as prices associated with reaching the allocation of capacity solution most efficiently.

It was remarked that the constants of the restraining equations in the minimiza-tion problem reappear in the associated maximization problem. In one case we are associating prices with traffic (via the a_{ij}'s) and in the second case we are associ-ating prices with resources (via the $b_{ij,\,kl}$'s and $c_{ij,\,kl}$'s). Put broadly, from the highway planner's point of view there are benefits that appear as tolls. From the resource planner's point of view there are benefits that appear at the resource use and consumption places. In the two preceding statements benefits are identified by the theoretical prices.

The discussion that preceded emphasized with some artificiality two points of view for highway development. To some extent these two points of view actually appear at the levels of policy and practice. The discussion has shown how the two points of view regarding the allocation of capacity are associated with two points of view regarding benefits. To the extent that the two points of view are legitimate, the two views of benefits are legitimate and would serve as legitimate guides for action.

There is always room for improving decision systems, and to the extent that an improved decision system is one that incorporates the two points of view ar-tificially exaggerated previously, the two points of view regarding benefits should appear in any broader scheme.

Summary

Analysis is nothing more or less than applying a point of view and a set of analyt-ic tools to a problem. In this volume the problem of an efficient transportation sys-tem is viewed as one embedded within larger problems of a society spatially or-dered by specialization and variations in resources. From this point of view, it was emphasized that developments of transportation may stem from opportuni-ties either for the more efficient operation of activities served by that transporta-tion or for opportunities for more efficiently carrying on the transportation pro-cess. In practice it is felt that combinations of both should enter development de-cisions, and the meaning of this on a policy level is that both vehicular and non-vehicular benefits of highway improvements should be considered. Others with other points of view might emphasize other benefit questions. Pricing the highway service to achieve an efficient division of traffic among rail, water, air, and highway transport is an example of a problem that might be stressed by other analysts.

We have several points of view which in a large measure are complementary. Supporting one does not mean that it is correct to the exclusion of others. Existence of several legitimate points of view emphasizes the need for a general point of view which encompasses these legitimate subparts. This chapter has presented fragments of a point of view of benefits from highway development.

SECTION II

Highways and Retail Business

This section comprises four chapters each of which contributes to an integrated discussion of the relations of spatial systems of retail business and the highway system. Chapter 3, entitled "The Spatial Organization of Business Land Uses," contains summaries of previous empirical work, planning concepts, and available theories. It shows how ordered spatial patterns of business land uses have been recognized and can be analyzed theoretically, and how available theory of business is compatible with theories of retailing, consumer behavior, and land uses. In Chapter 4, "Empirical Verification of Concepts of Spatial Structure," ordered patterns of business types are seen in a variety of empirical situations. One result of the chapter is to recognize "nucleated," "automobile row," "urban arterial," and "highway-oriented" business types based upon their locational and grouping habits. Changes in sales volumes of business establishments from a situation prior to, to a situation immediately after, the reorientation of an arterial highway in a small urban center are recorded in Chapter 5, "Impact of Highway Improvements on Business Establishments." Here groupings of businesses through time are noted in the same way that Chapter 4 recorded spatial groupings. Chapter 6, "The Relations of Highway Impact, Spatial Structure, and Urban Planning," attempts to integrate many of the approaches in the previous chapters. Movements of sales of spatial groups through the period of reorientation are studied. Present urban planning for business is evaluated, and possible effects of new locational opportunities created by the interstate system studied. In total the section attempts to establish the spatial structure, to record changes resulting from highway reorientation, and to view a variety of questions in light of these findings, all related to the general problem of the development of highway systems.

3

The Spatial Organization of Business Land Uses

DO BUSINESS land uses organize themselves spatially? Can ordered locational patterns of business land uses be associated with such a parameter as the availability of highway transportation? These are examples of questions to be considered in light of previous empirical work and available theory. Such consideration lays the groundwork for subsequent empirical studies reported in Chapter 4, and analyses of highway impact presented in Chapters 5 and 6.

It might be objected that a discussion of the present kind reveals nothing about the nature of influences of highway changes upon retail business. But to understand the nature of highway impact--to say *how* and then *how many* changes have come about because of changes in highway facilities--one must of necessity understand the fundamental spatial or geographical organization of the economic system. Without this understanding one cannot say *how*, much less estimate *how much*. It is to ease the work of saying *how* by clarifying the nature of the underlying spatial structure that the present discussion has been undertaken.

Initial sections of the chapter review a variety of previous empirical studies. The review indicates intuitive recognition of a spatially organized system of business land uses. Subsequent sections of the chapter tend to confirm this finding, for in discussions of available theory it is found that logical interpretations from the point of view of location economics may be given to the empirically observed regularities and to patterns suggested by planners. Other reviews of previous work are available, notably those of Canoyer (1946) and Kelley (1956). In contrast with these, the present discussion devotes much attention to central place theory and to the theory of tertiary activity. These, it is suggested, are statements of substantial power and generality. Power and generality of the theories are further evidenced by demonstration of their compatibility with modern theories of retailing and consumer behavior. In conclusion, relations of hypothesized patterns of business land uses with more general systems of land uses and rents are discussed. Here again, compatibility is revealed, and this tends to bear out notions of underlying lawful order in patterns of business land uses.

Observed Patterns

Several empirical studies of the spatial structure of retailing are available. This section considers the large scale pre-1940 studies of Rolph, Proudfoot, and Ratcliff; an immediate postwar review of these by Canoyer; a more recent study of a portion of the Chicago metropolitan area by Garrison using a classification of centers derived from the pre-1940 studies; and a study of a single string street in Denver by Merry. Details of these studies of particular concern here are: What spatial patterns and typology of business centers did the authors recognize? What types of business were shown to be characteristic of each type of center? What is an optimal or successful locational pattern for the various business types? Canoyer's review also addresses at length problems of retailing techniques and practices. These, however, are not of concern in the present context.

Rolph's Study of Baltimore

Using data collected in 1929 for the 1930 U.S. Census of Distribution, I. K. Rolph (1929) undertook a study of the locational structure of retail trade in the city of Baltimore for the Bureau of Foreign and Domestic Commerce of the U.S. Department of Commerce. Rolph concluded that the retail trade pattern of Baltimore could be described completely in terms of the following types of centers: the central business district, retail subcenters, string streets, neighborhood facility groups, nonconcentrated businesses.

In the central business district general merchandise, apparel, jewelry, furniture, and household facilities predominated. Of greatest significance in the retail subcenters were the food and apparel groups. String streets were observed to be highly specialized forms of development along traffic arteries, and facilities in the automobile, furniture and household, general merchandise, and apparel groups were most in evidence. Neighborhood facility groups and nonconcentrated facilities were characterized mainly by the food, automobile, lumber and building material groups. Detailed descriptions of number of businesses and sales volume by business groups for a variety of subcenter, string street, and neighborhood business centers were also provided.

Rolph's study is of importance because it was one of the first to show that various types of business centers exist, and that their functional make-up differs, even though the precise functional character of centers was masked by the highly aggregative groups of business types utilized.

Studies by Proudfoot

Based upon detailed studies of Chicago, Philadelphia, Cleveland, Atlanta, and Des Moines, and reconnaissance studies of Washington, New York, Baltimore, and Knoxville, M. J. Proudfoot (1937 etc.), a geographer working for the Bureau of Foreign and Domestic Commerce of the U.S. Department of Commerce, concluded that business centers within urban areas could be classified into five types: the central business district, outlying business centers, principal business thoroughfares, neighborhood business streets, and isolated store clusters. In addition there were noted, but not provided with a special type, the many isolated stores which exist in any city.

The central business district was recognized as the retail heart of the city, with a whole range of stores including the most specialized urban functions. Outlying business centers were considered miniatures of the central business district, located at focal points of the intercity transportation system and drawing customers from wide areas. Principal business thoroughfares were characterized by the coexistence of two related attributes: as business streets they possessed widely spaced convenience and shopping goods stores; as traffic arteries they carried mass vehicular traffic which "although the stores are primarily dependent on customers derived from the traffic have little countereffect on its density, but thrive by attracting small fractions of the passengers of this traffic." Isolated store clusters consisted of two or more complementary stores of the drug, grocery, meat, fruit and vegetable, and delicatessen family. Neighborhood business streets form a regular network throughout the city, with groceries, meat, fruit and vegetable, drug, and other convenience goods dominating.

One general conclusion of the studies was that "experiments have revealed that retail structures of cities of less than 250,000 inhabitants, for the most part, are centralized to such a degree that, aside from a few scattered outlying business streets and isolated stores, there is little business done outside the central business district."

From the Chicago study (1936-1938) Proudfoot ascertained that successful businesses tended to locate in the following manner:

Central business district--cigar, department, drug, jewelry, men's clothing, variety, women's clothing;

Outlying business center--apparel accessories, cigar, department, drug, fancy grocery, general merchandise, household appliance, jewelry, men's clothing, millinery, shoe, variety, women's clothing;

Principal business thoroughfare--auto accessories, tire and battery, farmer's produce, filling station, grocery, milk depot, new cars, used cars;

Neighborhood business street--bakery and catering, combination grocery and meat, drug, fancy grocery, grocery, meat market, variety.

For the city of Philadelphia (1937) quantitative information was presented concerning number and per cent of stores, sales volume, and so forth for a variety of outlying business centers and principal business thoroughfares. These data perhaps reflect the inconsistencies of the asserted functional bases of Proudfoot's admittedly largely morphological classification, for out of 30 outlying business centers, 4 have more than 50 per cent of sales in the food group, 3 between 50 and 70 per cent in the automotive group, while, similarly, of the 16 principal business thoroughfares 6 are dominated by sales in the food group, 4 by automotive establishments. Presumably, this great diversity suggests the poor quality of the functional bases of the classification system.

Ratcliff's Studies

R. U. Ratcliff has published the results of two major studies of business, one concerned with the outlying business structure of Detroit (1935, 1949) and one with associations of business types within central business districts (1939).

From the former study Ratcliff argues that the focal point of every city is the central business district, which has the most intense retail activity at the convergence of all traffic and transportation routes. Beyond this area the pattern of business structure consists of combinations and variations of two basic conformations. These are:

1. String street developments (business thoroughfares), which consist of businesses located along traffic arteries, but rarely down intersecting streets. The nature of uses along the string street depends upon the extent to which the street is a main traffic artery and the degree to which it is the core of a residential area. Uses serving the artery and uses intended for immediate convenience of local residential areas are both attracted to the artery. Residential functions appear as nucleated "beads" in string street developments.

2. Business nucleations, which consist of clusters of retail uses at important intersections, creating pyramiding of land values to peaks at the intersections. Nucleations vary in nature from isolated grocery and drug stores and neighborhood facility combinations to major retail subcenters and to the central business district, which is itself described by the pattern of nucleation. Smaller nucleations are usually concentrated on a primary street, though frequent spurs may extend down cross streets.

Retail affinities were studied by Ratcliff in his study of land uses in central business districts. The basis for the study was the frequency with which types of retail activity occur together in the same block. Suggestive of patterns of successful coexistence of retail establishments is Ratcliff's summary statement of affinities:

> Department stores--women's clothing (in 35 per cent of blocks in which there are department stores), shoe, variety, drug, men's furnishings, and millinery;
> Variety stores--women's clothing (60.7 per cent), shoe, department, jewelry, men's furnishings, and millinery;
> Shoe stores--weak affinity for women's clothing, department, variety, men's furnishings, restaurants;
> Jewelry stores--no strong affinities, most frequent neighbors are restaurants, shoe stores, and men's furnishings;
> Furniture stores--restaurants (40 per cent), men's furnishings, shoe, women's clothing;
> Florists--restaurants (47.5 per cent), women's clothing, drug, shoe stores;
> Theaters--restaurants (58.5 per cent), jewelry, men's furnishings, shoe stores;
> Women's clothing--variety (60.7 per cent), department and shoe stores;
> Millinery--women's clothing, variety, shoe, department stores;
> Furs--furniture stores, florists, shoe stores;
> Candy stores--theaters, jewelry, variety, florist;
> Men's furnishings--restaurants, furniture, theaters, men's shoes;
> Haberdashery--jewelry, men's furnishings;
> Tobacco--shoe, jewelry, theaters;
> Restaurants--florists, theaters, furniture, jewelry;
> Drugs--may appear near any stores except furniture (25 to 35 per cent of all central shopping district blocks contain drug stores);
> Curtain stores--near furniture stores;
> Paint stores--near furniture stores;
> Shoe repair stores--may appear near any stores except department or variety;
> Barber shops--near florists, theaters, jewelry stores;
> Grocery stores--seldom appear in central shopping district.

In this summary, Ratcliff provided the first detailed information concerning the importance of structuring of business locations internal to business districts as well as between alternate business centers.

Canoyer's Review

H. G. Canoyer, of the Marketing Division of the Office of Domestic Commerce, U.S. Department of Commerce, reviewed many of the prewar studies of locational structure of retail trade in order to establish criteria for successful selection of locations by retail businesses (1946). Canoyer emphasized the critical importance of good location, writing:

> The location of a retail store plays a vital part in its success. Regardless of the size of the store or the kinds of merchandise offered for sale, the location must be suitable or sales volume will suffer, profits will be restricted, and failure may be the ultimate result. Although good locations frequently offset deficiencies in retailing, poor locations seriously handicap the most skillful merchandisers.
>
> Studies which have been made of retail store failures reveal that a poor location is often associated with failure. In a study made in Massachusetts it was found that 28 per cent of the retail stores which failed had poor locations. A National Drug Store Survey revealed that, of those stores which failed in St. Louis in 1932, the majority did not have good locations and one-third had chosen sites where drug stores had previously failed.
>
> The problem of a good location is as vital a one to the man who is already in business as to the prospective retailer. Because population centers do not remain constant and the value of a site fluctuates, the problem of choosing a good location is a continuing one. A market area is a dynamic thing, constantly adjusting to the economic and social forces which determine its pattern. Therefore, the retailer must be alert to trends and anticipate significant shifts in the shopping centers. A good location of today may, in a few years, become a poor one.

This most significant conclusion which Canoyer derived from her review of previous work is just as vital today, especially when many system-wide changes in economic structure are to be expected from improvements in the Federal Highway System.

Canoyer suggested that the shopping center structure which could be recognized as a result of previous studies was one consisting of *string streets,* with narrow-line shopping stores located with respect to high vehicular traffic and relatively low pedestrian traffic and with patronage largely derived from users of the thoroughfare rather than from nearby residents, and a variety of *cluster types* of shopping centers. The cluster types were classified as the central business district (where retail uses predominate at the point of maximum convergence of traffic flow in the city and all types of goods are sold, with pedestrian traffic at a maximum) and two lower orders of center. The major of these two lower orders of center is the major subcenter, a community shopping district, and the minor is the neighborhood center in which business is restricted to convenience and service establishments wholly dependent upon the immediate neighborhood for support. In addition to string street and cluster types, a variety of isolated store locations were noted. Canoyer concluded that isolated stores either were of such a specialized character that they were able to draw trade to their isolated site with little difficulty or else they had to engage in widespread advertising to survive.

Garrison's Study in Chicago

W. L. Garrison (1950) undertook a study of the business structure of a portion of the Chicago Metropolitan Area, the tributary area of the Fountain Square Major Outlying Business Center. In this study, Garrison took a classification system of business centers evolved in prewar empirical studies of business (community business centers, major and minor neighborhood business centers, principal business thoroughfares, isolated business establishments) and provided detailed information concerning the size and spacing of centers, and the types of business characteristic of each type of center.

Twelve community business centers were studied. These averaged 144 establishments, each representing 53 business types. In them particular concentrations of food, cleaning and laundry, eating, medical, apparel, and fuel and automobile facilities were noted. All twelve centers were located central to mass transportation facilities, and they suffered from vehicular and pedestrian congestion at rush hours. In densely populated apartment house areas they were a half mile apart, in other areas of continuous residences one mile, and in less densely populated areas of fragmented settlement they were one and a half miles apart. This spacing was not affected by location relative to the Fountain Square Major Center. They possessed an average of only one-quarter the business frontage of the Fountain Square Center.

In thirteen major neighborhood business centers studied, food, fuel, eating, cleaning and laundry, furniture, household, and radio sales, automobile service, beauty and barber facilities were most evident. These centers were located either at centers of mass transport facilities or on local bus routes and principal thoroughfares. They had but little problem of traffic congestion at any time, and they possessed an average of only one-third the business frontage of the community centers, or 31 establishments representing 19 kinds of business. Confined to areas of moderately high population density, their spacing was affected by the location of larger centers, but they were only slightly further removed from larger centers than from each other. Average spacing was one-third of a mile.

Duplication of establishments of the same type was virtually absent from the 25 minor neighborhood business centers studied, for they possessed an average of nine establishments representing eight types of business and had only one-quarter the business frontage of major neighborhood centers. These minor neighborhood centers were confined to local bus routes or local access streets and were often spaced as little as a block apart. In them food, eating, fuel, beauty, barber, cleaning, and auto service establishments were most evident.

On the 21 business thoroughfares were located stores which catered largely to preselected wants: fuel, automobile services, eating, building materials, recreational services, auto dealer, beverage sales. Most of these required large and expansive sites, and sold articles which are needed infrequently or by restricted numbers of individuals, or of such a sort that shopping is no prerequisite for buying. These thoroughfares averaged 15 establishments representing 9 business types.

Three types of orientation, each with differing functions, were noted among the 255 isolated business establishments. Located on principal thoroughfares were gas, restaurant, beverage, or automobile service facilities. Food stores were usually located with respect to local access streets. Isolated fuel dealers and such noxious activities as ice stores were oriented to railroads. Also isolated were building material, moving, and storage facilities.

Merry's Study of a String Retail Development in Denver

A study by Paul R. Merry (1955) was devoted to the analysis of a single string retail development: East Colfax Avenue, Denver, Colorado. Main features of significance in this work were a careful description of the business types locating along the highway, an analysis of differential location habits of the various types of business, and an outline of the relationships of the patterns described to such factors as traffic flow and numbers of people residing in areas immediately peripheral to the string development. A brief historical account of the growth of the development was also provided.

Activities found most frequently, by numbers, along the arterial were: restaurant, cleaner and laundry, auto sales, gas, beauty, drug, barber, and grocery. By ground area occupied, the most significant uses were: auto sales, gas, grocery, restaurant, drug, cleaner and laundry, auto repair, and furniture. Of these business types the auto sales and gas facilities were found most frequently on the north side of the arterial, and the beauty, baker, grocery, tavern, hardware and paint outlets were found most frequently on the south side. It was noted that westbound traffic (north side) was denser than eastbound, and contained greater proportions of its origins and destinations external to the city. Population numbers were greater in the three-block strip peripheral to the south side of the development than they were in the strip peripheral to the north side.

Generally, the density of stores diminished from west to east with increasing distance from the city center. However, there were marked variations in the pattern, with peaking of density at major intersections. Activities tending to agglomerate in high density zones (or nucleations) were: shoe repair, beauty, barber, grocery (independent), clothing (women's, infants' and children's, men's), shoe, hardware and paint, florist, millinery, liquor, bar and tavern, confection, delicatessen, poultry and meat, dry goods. Low density activities were: creamery and ice cream, auto sales, drug, gas, auto repair, grocery (chain). Cleaner and laundry, restaurant, furniture and household facilities showed no tendency to concentrate. Corner locations were favored by gas, grocery (chain), drug, and auto sales facilities more than 70 per cent of the time, whereas radio repair, jewelry, shoe repair, barber, grocery (independent), bakery, hardware, beauty, cleaner and laundry, photography, florist, delicatessen, and appliance facilities were found with similar or greater frequency within blocks.

Suggestions from Marketing and Planning

Much interest in locations and types of retail developments has also been evidenced in the fields of marketing and planning. At an early date marketing experts saw fit to distinguish between various types of goods marketed, dividing them into three classes: "convenience, " "shopping, " and "specialty. " Meanings of these terms are, of course, widely known (see for example the standard early description by Copeland).

Intuitively, at least, associations of store types with the three classes of goods have been made. Such associations have become standard in the fields of planning and shopping center development, and based upon the associations, classes of shopping centers have been identified. A variety of definitions is available for each of the classes, as Kelley (1956) has shown. This variety reflects, perhaps, the intuitive character of the underlying notions and their lack of empirical foundation. Descriptions of the classes of centers follow.

Neighborhood Centers

The lowest order of center recognized is the neighborhood center, for in the available literature any aggregations of business comprising fewer than 6 stores were not accorded the rank of shopping center. Various alternate definitions of the neighborhood center are:

1. Serving a minimum of 750 families, always with a small supermarket and drug store, dry cleaner and laundry, beauty, barber, shoe repair, and variety (Baker and Funaro, 1951);
2. Serving 500 families, with 10-12 stores supplying convenience goods. The first 10 stores should include drug with some eating facilities, cash and carry grocery, cleaner with laundry agency, beauty, filling station, bakery, shoe repair, barber (Community Builders' Council, Urban Land Institute, 1948);
3. Serving 10, 000-20, 000 people, with a core of food, drug, and other stores dispensing convenience goods (Gruen and Smith, 1952);
4. Serving 1, 000 families, with 10-15 stores dispensing convenience goods, including food, drug, sundries, and personal services (McKeever, 1953). In 19 shopping centers studied, stores occurring 10 or more times were: drug, supermarket, variety, bakery, dry cleaner, beauty, women's clothing, shoe, gift stationery and books, babies' and children's clothes, jewelry and watch repair, barber, men's wear, candy, shoe repair;
5. Potter (1943), president of the Urban Land Institute, suggested the pattern in Table 3-I for neighborhood facilities;
6. V. L. Lung (1955) provided data concerning the occurrence of store types in eight Seattle neighborhood centers. Functions which occurred 50 per cent of the time or more were: grocery, beauty, barber, cafe, tavern, service station, cleaner, drug, hardware, realty, variety, bakery, apparel, gift. Lung also summarized much of the work of the studies outlined above (1-5), and concluded that

    a. grocery or supermarket, drug, service station, cleaner, barber, beauty *always* occur in neighborhood centers;
    b. variety, hardware, shoe repair, real estate, bakery, laundry, restaurant, delicatessen *frequently* are found in neighborhood centers;

TABLE 3-I
NEIGHBORHOOD SHOPPING CENTERS
(50 to 2,500 Families, Average Per Family Estimated Income: $2,500)

Number of Families	Available Shopping Money	Type of Store
50	$ 55,900	General store
250	279,500	Supermarket Drug store Bar and grill
500	634,000	Supermarket Drug store Stationery Bar and grill Dry cleaner Laundry
1,000	1,268,000	Supermarket Drug store Stationery Bar and grill Delicatessen Bakery Beauty parlor Dry cleaner Laundry
2,500	2,170,000	Supermarket-2 Drug store-2 Stationery-2 Restaurant Delicatessen Bakery Beauty parlor Dry cleaner Laundry-2 Hardware Florist Barber shop Liquor store Bowling alley with bar and grill

c. professional and medical office, women's apparel, men's clothing, furniture and appliance, TV and radio sales and service, tavern, gift, florist, jewelry, liquor *sometimes* are located in neighborhood centers;

d. bank, fix-it, candy, bowling alley, sporting goods, post office, theater are *rarely* located in neighborhood centers.

The variety of definitions of neighborhood centers is readily evident from the above.

Community Centers

Community shopping centers are the second order of shopping facility. Again a variety of alternate definitions is to be found:

1. In addition to neighborhood facilities, have radio-TV, children's and specialty clothing, gift, candy, restaurant, haberdashery, florist, women's clothing, and theater in the older unplanned types. These centers serve 20,000 to 100,000 people and have as their core a large supermarket or a small department store (Kelley, 1956);
2. Serve 15,000-30,000 people, with 16-35 stores, and a complete range of convenience, shopping, and specialty goods, emphasizing apparel and home furnishing, professional office, and a bank (Baker and Funaro, 1951);
3. Community (or suburban) centers, in addition to convenience goods, have such hard and soft lines as apparel and hardware. Such centers contain 20-40 stores and serve 5,000 families (McKeever, 1953);
4. Serve 1,000-3,000 families and possess 25-40 stores. In addition to basic neighborhood facilities, such stores should include (second ten) service grocery, florist, milliner, radio and electrical shop, 5 and 10 cent store, children's shoe and shoe repair, gift, candy and nut, lingerie and hosiery, and liquor; (third ten) fix-it, dress shop, theater, frozen foods, cafe and drive-in restaurant, book and stationery, dentist and physician, baby goods and toy, haberdashery, athletic goods (Community Builders' Council, 1948).

Suburban Centers

McKeever (1953) considered that community and suburban centers were of the same type, but Kelley (1956) sought to distinguish between them, describing suburban centers as serving 30,000 to 100,000 people, built around a large department store and several large supermarkets. Except for unusual specialty items, all facilities are found, and where such centers are not planned they function as miniature central business districts.

Regional Centers

Several definitions of regional centers are also available.

1. Serve 100,000-1,000,000 people residing within 30 minutes' driving time of the center. Contain 1-2 major department stores and all convenience, shopping, and specialty goods. Always planned (Kelley, 1956);
2. Serve more than 50,000 people, and have as their basis department, apparel, household, and appliance facilities and theater (Hoyt, 1949);
3. Serve more than 100,000 people and contain 50-100 stores providing complete one-stop shopping. Always include home furnishing facilities, and compete vigorously with the central business district (McKeever, 1953).

The Planning View Concerning String Streets

It has been common in planning circles to label the string street type of development as "uneconomical and hazardous" (Courtney, 1955). One student writes, for example, that

> . . . there is little to support the retention of commercial development along the highway. Consolidation of urban facilities is urgently needed to restore stability to property values and convenience for the daily multitude who daily patronize the variety of business enterprise in our cities. It is also needed to stem the insidious spread of blight which is gnawing at the core of cities. This consolidation will come by recreation of "centers" of business connected but not traversed by traffic arteries and the rejection of "strip" or "shoestring" zoning along the highways which retain the horse and buggy tempo of the village [Gallion, 1950].

Summary of Previous Works

It is evident that there has been but little *explicit* agreement upon questions relating to the nature and structure of urban business centers. This is particularly true because much of the work has been quite intuitive, as is evidenced by urban planning and marketing research. However, some summary of the pattern of urban business as described in previous studies may be attempted, if only to form a reference point for discussion below of theories of urban land use.

Most previous work recognizes two types of centers, nucleated and string street. Empirical studies simply record the fact of their existence; city planners look for the elimination of the string street type. Nucleated centers include the central business district, regional (major outlying) centers, community (including or excluding suburban) centers, and neighborhood centers, in order of size. String streets (principal business thoroughfares) vary in size but are not generally differentiated in the same manner as nucleated centers. Kelley (1956) considered that nucleated types should also be subdivided into planned (controlled) and unplanned centers. All studies also recognize a variety of isolated stores and store clusters, and Garrison (1950) differentiated these into three types according to their locational orientation. Reference to the descriptions of studies above will provide some understanding of the store types found or asserted to be most successful in each type of center. It is generally assumed that the stores characteristic of larger centers, but not of smaller, are those which require larger market areas to be economic undertakings. Observed differences between business types characterizing nucleated and string street centers were asserted to result from the emphasis upon service of traffic in the latter, and service of the consumer making specific shopping trips in the former.

Factors Causative of Observed Patterns

This, of course, raises questions concerning factors suggested to be causative of observed locational patterns of urban business. A variety of factors has been proposed. One group comprises population distribution, income levels, and available purchasing power within the urban area. Differences in the incidence of purchasing power are held to be the cause of differences between functions located in nucleated and string street centers. A second set of factors relates the location of centers to the pattern of transport and traffic facilities within cities, and to both directional flow and orientation of traffic; this set of factors can be subsumed under the general definition of accessibility. Third, characteristics of business establishments have been mentioned, particularly the influence of the threshold market factor (the minimum size of market necessary to support a profitable establishment). Consideration of the supplying firm suggests a fourth factor: location and characteristics of potential competitors. Selection of a site within a business center by a business establishment has been attributed to two factors: whether the firm is willing to pay high or low rents, and what particular affinities to other stores are of greatest importance. Other factors mentioned include available buildings and plant, and parking facilities of the center. Canoyer (1946) discusses in more detail many of these suggested factors.

Theoretical Approaches

Listing of causative factors is not enough. A variety of questions remains unsolved. Can the listed factors generate a business structure comprising basic nucleated types (of various orders) in addition to string street types of business centers? Are there other important causal processes? What are the relations of the factors to each other, and exactly how does each operate? What are the effects of a change in any one of the operative causative factors? What relations does the structure of business have with the more general systems of land uses and rents? What are the relations of the structure to patterns of movement and to the highway system?

What is needed is to gain an understanding of how the systematic structure of land uses obtains, given the many listed factors as locational variables. This need is addressed in the following section. First, certain basic theoretical notions from central place theory are presented and then adapted to the case of urban business. Second, the various factors are evaluated critically in light of the theoretical system, and interpretations of the spatial structure observed or hypothesized in previous work are given. Third, compatibility of the theory of business land uses with modern theories of imperfectly competitive retailing and consumer behavior is discussed. Finally, relations of the systematic structure of business land uses and the system of urban rents and land values are considered.

Central Place Theory

Originally, central place theory was developed by German geographer Walther Christaller (1933) and German location economist August Lösch (1944) to explain the size, spacing, and functions of urban centers. It is not practicable to reproduce all the details of their theories here. For these the reader should refer to the original sources, or to such reviews as those of Berry (1956), Ullman (1941), or Berry and Garrison (1958 a, b, c). In the following sections, an outline of the contributions of Christaller and Lösch will be provided, and a recent, elegant, concise summary statement by Beckmann will then be utilized as a device to facilitate immediate appreciation of relationships between important causative variables.

Christaller

Central place theory as formulated by Christaller in his *Die zentralen Orte in*

Süddeutschland (1933) and presented elsewhere by Ullman (1941) is relatively well known. The content of the theory may be stated in a summary way with an outline of its definitions, relationships, and consequences.

A. Terms defined:
   1. A central place.
   2. A central good.
   3. A complementary region.
B. Relationships specified:
   1. Variations in prices of central goods as distance from the point of supply changes.
   2. Explicit extremization behavior in the distribution and consumption of goods (e.g., goods are purchased from the closest place).
   3. Inner and outer limits for the range of distances over which central goods may be sold.
   4. Relationships between the number of goods sold from a central place and the population of that place.
C. A statement which used the terms defined and relationships specified (within the simplifying assumption of homogeneous distribution of purchasing power in all areas) and described the arrangement of central places and complementary regions.
   Essential features of this statement were:
   1. Hexagonal market areas for any set of central goods.
   2. Overlapping sets of hexagons. The hexagons overlap in such a way that larger hexagonal market areas (resulting from a set of central goods) are divided into smaller hexagons when supplied by other central goods. The smaller hexagons nest into the larger according to a rule of threes (this is the k=3 network described by Lösch).
   3. Transportation routes serving the system of cities.

Lösch

A good portion of Lösch's now classic *Die räumliche Ordnung der Wirtschaft* (1944) was given to an evaluation and extension of notions of the arrangement of central places. Lösch's additions to the notions fall into three major divisions.

A. Explicit statement of two aspects of the system:
   1. The derivation of demand cones over areas for goods.
   2. Verification of the hexagonal-shaped complementary region as the "best" shape where purchasing power was uniformly distributed.
B. Clearly linking the arrangement of transportation routes among cities to central place notions.
C. The extension of the special case of a k=3 network to a more general description of a system containing all possible relationships of evenly spaced central places and nests of hexagonal shapes of complementary regions. In the system he develops, he maintains:
   1. That consumer movement must be minimized.
   2. That no excess profits can be earned by any firm.

Lösch further asserts, but does not demonstrate, that one outcome of his system is a hierarchical class system of central places both as sources of central goods and as intersections upon the transport network.

Beckmann's Summary Model

Beckmann (1958) developed the simplest model of central place theory in the following manner. There is a basic level of rural population which is served by a first layer of settlements performing the most elementary production and distribution functions. Market areas of these first settlements are limited by the largest distance tolerable to the rural population served. A fundamental hypothesis is that the size of any city is proportional to the population it serves (including that of the city itself). Therefore, if r is the rural population in the market area of cities of the first layer, k the proportionality factor, and c the city size, then

$$(1) \qquad c = k\,(r + c), \qquad \text{or}$$

$$(2) \qquad c = \frac{kr}{1-k}$$

and k/1-k is therefore the effective "urban multiplier."

It is assumed that cities of each order (but the lowest) have a fixed number of satellites, s, of the next lower order; it is now possible to derive the size p_m of cities of each order m, and the population P_m served by these.

From the population level hypothesis $p_m = kP_m$ and from the assumption concerning satellites, $P_m = p_m + sP_{m-1}$. By substitution (see Beckmann) it may be shown that

$$(3) \qquad P_m = \frac{s^{m-1}r}{(1-k)^m}$$

and

$$(4) \qquad P_m = \frac{ks^{m-1}r}{(1-k)^m}$$

where r is the size of the basic rural community, k is the ratio of city size to population served, and s is the number of satellites per city.

When m assumes its largest value, say N, then (3) describes a relationship between population P, the size of the smallest community r, and the number of ranks in the city hierarchy N:

$$(5) \qquad P = P_N = \frac{s^{N-1}r}{(1-k)N}$$

This may be written in other ways:

$$r = \frac{(1-k)^N}{s^{N-1}}P$$

$$N = \frac{\log\,(Ps/r)}{\log\,[S \div (1-k)]}$$

A second relationship exists between the total number of cities T, the number of ranks in the hierarchy N, and the number s of satellites per city:

$$(6) \qquad T = 1 + s + s^2 + \ldots + s^N = \frac{s^{N+1} - 1}{s - 1}$$

To summarize, there are six variables: k, s, r, P, N, T. If certain of them are given, the others are determined by the model.

Burros' "Control Ratio" Concept

R. H. Burros (1957) presented recently a theory of static hierarchy which facilitates the generalization of certain aspects of Beckmann's statement. In the Beckmann model there is a single ratio (s) which describes the number of satellites per city. Burros recognizes that this *control ratio* (R in his system) may have several values between lower and upper bounds, and his model makes explicit recognition of this fact. In the Burros model the variables are:

E, the total number of echelons in the regular hierarchy 1, . . ., E
C_i, the number of centers at echelon i
C_e, (or S) the number of centers at the lowest echelon E
L, the total number of centers at all echelons
R, the control ratio, the number of centers at echelon i + 1

controlled by centers at echelon i.

$R_i \leqslant C_{i+1} / C_i \leqslant R_2$ and R_1 and R_2 are the lower and upper bounds of the control ratio, respectively.

Generally, $C_1 = 1$ and $C_E = S \geqslant 2$

$$1 < R_1 \leqslant C_i + 1 / C_i \leqslant R_2$$

$$L = \sum_{i=1}^{E} C_i$$

It may be proved that

$$(1) \qquad E_j = 1 + \log S / \log R_{3-j} \qquad j = 1, 2$$

i.e., $E_1 = 1 + \log S / \log R_2$

$\qquad E_2 = 1 + \log S / \log R_1$

$$(2) \qquad L_j = (SR_{3-j}-1) / (R_{3-j}-1) \qquad j = 1, 2$$

i.e., $L_1 = (SR_2-1) / (R_2-1)$

$\qquad L_2 = (SR_1-1) / (R_1-1)$

By applying (1) and (2) it is possible to obtain the range of possible values of E and L. From this one can calculate and draw possible regular hierarchies. Note that R_1 and R_2 relate to the "span of control" in hierarchies, and provide a more flexible definition of the s in the Beckmann model.

The Theory of Tertiary Activity

As developed, central place theory related only to alternate urban centers and the transport network linking urban centers, although in the Burros case relations with organization theory are apparent. The statement was in terms of homogeneously distributed purchasing power (hence, the hexagonal trade areas) and, thus, was restricted in applicability in more realistic situations. Also, such assumptions as the absence of excess profits in the system were troublesome. Since Lösch's contributions, however, theory has been reformulated to consider not only tertiary activities in alternate urban centers but also retail and service business of shopping centers within cities. Central place theory has thus been extended to a general theory of the spatial structure of tertiary activity. The extended theory is outlined below. Note that (a) a hierarchical structure is developed without the uniformity assumptions concerning purchasing power, and (b) the notion of absence of excess profits is relaxed; and the theory is therefore applicable in imperfectly competitive situations as well, with some excess of capacity over the perfectly competitive minimal norm.

There are two definitions necessary for the formulation of the theory of tertiary activity:

1. The range of a good. The range of a good offered from a business center has an upper limit beyond which the business center is unable to attract purchasers for the good and a lower limit which incorporates the threshold purchasing power needed for the good to be offered (see below). It thus defines the market area of the business center for the good.
2. Threshold. There is some minimum size of market below which a business center will have no economic justification to supply a good. This is the point at which sales are large enough only for the firm to earn normal profits. It constitutes a minimum scale (or condition of entry) defined by the lower limit to the range of a business center for the good. Such a minimum scale of purchasing power necessary to support the supply of a central good from a business center is here termed the threshold sales level for the provision of the good from a center.

For the sake of exposition, assume that a city is to be supplied with n types of central goods. Let these be ranked from 1 to n in ascending order of their threshold sales requirements. The business center supplying good n will require the largest market area (in terms of amount of purchasing power) for its support. Let a central place supplying good n be called an A center.

As many A centers will exist in the city as there are threshold sales levels to support firms supplying good n. These firms compete spatially, hence are distributed so as to supply their own threshold most efficiently. If total sales levels are an exact multiple of thresholds for good n, market areas will be bounded by lower limits to the range of A centers. Firms will earn only normal profits and these only if they minimize costs by locating to minimize distribution costs if the product is delivered, or locating to minimize consumer movement if the consumer comes to purchase the product. Hence the extreme importance of accessibility in the system. If sales in the whole area are slightly greater than an exact multiple of threshold, but not great enough to justify another A center, than excess profits may be earned and ranges reach a more competitive upper limit.

The question arises as to how good n-1 will be provided. Presumably, it will be supplied from A centers, which have sought out the most efficient least cost points of supply central to maximum profit areas at their command. Also, there will be advantages of association with other establishments providing central goods. The threshold of good n-1 is less than that of good n and spatial competition determines market areas, which are delimited by upper limit ranges. Excess profits may be earned. This argument will be the case for all other goods n-2, n-3, . . ., 1.

But there may be one or more goods, say good n-i, in which case the interstitial purchasing power, located between threshold market areas of A centers supplying good n-i, will reach threshold size. In this case greater efficiency is reached if a second set of centers, which may be termed B centers, supply the good. These B centers again locate most efficiently relative to their threshold market areas. If the market area is just at threshold, only normal profits are earned by firms supplying good n-i. If part-multiples of threshold are present, some excess profits are earned. Good n-i may be termed a hierarchical marginal good. B centers will also provide lower threshold goods n-(i + 1), . . ., 1.

Let it be assumed that good n-j (j>i) is also a hierarchical marginal good, supporting a third set of central places designated as C places. These are a lower order of business center and provide only goods n-j, . . ., 1.

TABLE 3-II
THE SUPPLY OF n GOODS BY BUSINESS CENTERS

Center	Goods			
	n*, n-1, . . .	n-i*, n-(i+1), . . .	n-j*, n-(j+1), . . .	. . .
A	x	x	x	. . .
B		x	x	. . .
C			x	. . .
.				. . .
.				
.				

* Indicates a hierarchical marginal good.
x Indicates the set of goods supplied by the center.

The pattern of provision of goods by centers in this hierarchical system may be displayed in an array as in Table 3-II. The table displays how sets of goods build up hierarchies of types of business centers. For example, a set of C centers and places in the tributary areas of C centers rely upon either B or A places for goods n-i through n-(j-1) and upon A places for goods n through n-(i-1). B places rely upon A places for goods n through n-(i-1). All places are positioned at the point of most efficient supply of their tributary areas.

Excess profits may be earned in the system. Where n goods are provided, it

is likely that the hierarchical marginal firm will tend to earn only normal profits. This is the firm which satisfies Lösch's conditions that all excess profits
shall be at a minimum. However, all supramarginal firms in the hierarchy will
have an opportunity to earn excess profits to the extent that they are able to compete spatially with other firms for purchasing power located between threshold
market areas in the spatial system, and, therefore, excess capacity over the perfectly competitive norm will exist in this case of imperfect spatial competition.

Brink and de Cani's Model

Brink and de Cani (1957) have provided an analogue solution to the problem of
locating retail points most efficiently (i.e., in a least cost manner) relative to
market areas. The solution is also applicable to wholesale-retail relationships.
With a little practice it is possible, using this technique, to derive optimal solutions for small scale problems by hand with the assistance of little more than a
desk calculator. Presentation here is motivated by the obviously valuable relationships to the theory of tertiary activity presented above.

The variables in the Brink and de Cani model are as follows:

y_j, a vector representing distribution point j

$y_j = (y_{j1}, y_{j2}, \ldots, y_{jr})$ and $i, j = 1, 2, \ldots, n$

$x_i(j)$, a set of vectors representing consumers i within the market area of y_j

$x_i = (x_{i1}, x_{i2}, \ldots, x_{ir})$

$T_i(j)$, the total demand of $x_i(j)$

$W_{ij} = T_i(j) / \{x_i(j) - y_j\}$

M_j, the set $\{x_i^{(j)}\}$ of n_j consumer locations for each y_j, a uniquely determined market

$\{x_i(j) - y_j\}$ are the costs of transport from j to i in the market area of y_j

From the variables it is shown (see Brink and de Cani) that

$$(1) \quad y_i = \sum_{i=1}^{n_j} W_{ij} x_i^{(j)} \Big/ \sum_{i=1}^{n_j} W_{ij}$$

$$(2) \quad \{x_i^{(j)} - y_j\} < \{x_i^{(j)} - y_k\}$$

Condition (2) is the nearest neighbor relation for markets M_j and M_k, uniquely
distributed. By alternatively meeting conditions (1) and (2), beginning by arbitrarily locating y_j in cost space, an optimal solution may be iterated. To map
back into geographic space from the cost space in which the optimal solution was
achieved, the steps are: (i) locate y_i at the $\{x_i(j)\}$ for which the distance is least;
(ii) define markets on the latter sets; (iii) sum over the market sets to obtain total
demands. This also gives total inventory at the y_j.

Suggested Factors and the Hierarchical Structure

Several types of factors were suggested previously to affect competitive bidding
for the land by businesses. The role of these factors may now be clarified within
the theory of tertiary activity developed above.

Accessibility

The basic importance of accessibility in the system has already been outlined. The locational choice of firms is intimately related to minimization of costs of consumer movement or costs of distribution, given the nature of the market area, for the successful firm locates at the least cost point central to the maximum profit area at its command. Hence, spatial patterns assumed by the hierarchy are obviously closely linked to patterns of movement and traffic flow within the city, for these reflect lines of least resistance or greater accessibility.

Characteristics of the Urban Market

The spatial distribution of central places of each rank is critically conditioned by the distribution of purchasing power. Where purchasing power is available in large amounts because of the concentration of consumers within small areas, centers of each rank are closely spaced; where purchasing power is widely dispersed, so are central places. Hence the differences in the distribution of central places which appear in the urban-rural continuum: cities have closely spaced centers; in rural areas they occur less frequently. The urban market comprises far more than a simple distribution of purchasing power, however. It consists of people with available purchasing power varying widely with income differences and, hence, with demands for baskets of goods of widely varying character; it consists of people of different social and ethnic backgrounds who have differential satisfaction functions and preference structures. Variations of this character are especially likely to create variations in the elements of the hierarchical structure. Certain goods may be provided in low income areas, and other goods are substituted for these in areas of higher income; kosher meat shops in a Jewish community may be substituted for the *Wurst-haus* of the German. Hence, the structure is viable in terms of variable characteristics of the urban market; their influences within the process-system may be traced.

Competition Between Stores

The theoretical structure shows that competition between stores should be interpreted in a special sense. In the spatial system, competition is for demands in excess of threshold located between threshold market areas. Competition is for these surplus demands over which the store does not have a spatial monopoly in terms of accessibility. It is also possible for competition to assume all the forms of monopolistic competition as well.

Characteristics of the Supplying Firm

Characteristics of the supplying firm are defined in terms of threshold requirements of the firm, and their role is quite explicit within the theoretical system.

Types of Activities in the Immediate Area

The hierarchical system contains definitions of complementary uses at each level of the hierarchy. Effects of complementary and antipathetic uses upon rent functions are discussed below.

Nucleations and String Streets

From previous empirical studies and work in marketing and planning it was suggested that there are two basic business types in the city, nucleated and string street. It was also remarked that various levels of the nucleated type could be seen in a regular hierarchy from the central business district to the neighborhood center. A variety of isolated store clusters was also noted. Does the theory of tertiary activity result in classes of centers of these types?

Nucleated Centers

A hierarchical structuring of business centers into several ranks or levels, from the single most complex center through increasing numbers at intermediate levels to largest numbers of centers at lowest levels, is postulated by the theory of tertiary activity. Higher level centers perform the functions characteristic of lower level centers, plus some additional functions which they perform for the service areas of the set of lower level centers falling within their "span of control." The most complex center performs specialized functions for the entire urban area. Individual businesses nucleate in centers appropriate to their rank in order to minimize costs of shopping by the visiting consumer, who generally comes from home, parks his car, shops on foot in the center (or walks from home in the case of the lowest level centers). These notions are, of course, directly translatable into the patterns recognized by previous students, and are also compatible with the patterns suggested in the fields of marketing and planning.

String Street Types

Conversely, some mental gymnastics are required to reproduce string street type centers from the theory. First, it must be recognized that string street type centers serve different demands than do nucleated centers, demands not for immediate consumer goods purchased from the home, but presumably demands associated with people moving along major urban arterials. String street type centers show clear relations to the pattern of traffic flow in the city, and it may be inferred that they serve people on the move (i.e., satisfying auto-based demands). Spatial distribution of demands in this case is not approximated by the population or housing map, as in the case of nucleated centers, but by the traffic flow map. Also, shopping is not by people walking to several stores in a center but is of a single-purpose character from an immediately parked automobile. Hence, businesses need not nucleate in centers to minimize the costs of shopping; they can string along major traffic arteries and still be located central to the maximum profit areas at their command. With these special limitations the presentation of the theory of tertiary activity above can be used to approximate the structure of string street business types.

Compatibility with Other Theories

We have tended to show so far that the theory of tertiary activity, a reformulation of central place theory free of assumptions concerning absence of excess profits and also assumptions concerning the shape and character of trade areas, provides a "good" explanation of patterns of business centers observed in previous empirical studies and hypothesized in marketing and planning. That is, the new formulation tends to meet requirements of correspondence with reality. Another asset of the new formulation is its relation to alternate theories. Certain of these relations are explored below.

Theory of Retailing

At first glance it appears that the theory of tertiary activity is compatible with the existing body of theory concerning the retail firm, as developed by such students as Aubert-Krier (1949), Smithies (1939), and Lewis (1945), since apparently the theory posits single types of business or essentially single-product firms which reach a competitive state of spatial equilibrium.

The lack of sophistication of such a single-product approach has been pointed out by Holton (1957). He has argued that a far more realistic theory of retailing could be developed within the framework of a multiproduct firm, and he has developed a model for this case. His conclusion is that the long-run equilibrium pattern of the multiproduct firm is one which requires that all products in which marginal revenue exceeds marginal cost be added to the product line, and sales of each expanded to the point where marginal profits are all zero. Empirical tests in the case of supermarkets have verified the conclusions of his model and the further finding that profit maximization will result in price discrimination, since products will face demand functions of different elasticities.

Theory of tertiary activity is in fact compatible with this formulation of the equilibrium of the retail firm. Consider each rank of the hierarchy of central functions as a firm. Many products will be supplied by this rank (the number of stores supplying them will, of course, be substantially less). All products will be sold for which marginal revenue exceeds marginal cost, and marginal cost is defined upon the threshold sales volume of the hierarchical marginal product. Sales are expanded to the point where marginal profits are all zero as determined by the upper limits of the range of each product.

Generally, stores will be located in a manner which minimizes consumer movement and hence will maximize profits. But for all products other than those of the marginal hierarchical goods, it is possible to earn excess profits because excess capacity exists as a result of existence of unequal thresholds of firms. The theory posits that these excess profits are allocated by spatial competition. It is obvious that notions concerning spatial competition should be broadened to include competition through the practice of price discrimination according to the elasticity of demand for different products. Ability to compete in these more general terms means that excess profits may be allocated in far from an equitable manner among existing firms and that some excess capacity may exist.

Notions of Consumer Behavior

It is interesting to consider correspondence between theory of tertiary activity and recent work regarding shopping behavior by customers. Baumol and Ide (1956) have produced a simple model displaying the choice variables in the selection of a shopping center by a consumer. A customer will shop at a center when his demand function is such that

$$f(N, D) = wp(N) - v(C_d D + C_n \sqrt{N} + C_i)$$

is positive. $f(N, D)$ is a measure of the expected net benefit of the consumer from entering a store. It varies with D, his distance from the store, and N, the number of items offered for sale at the store. Assumed costs are C_d, a cost of transport assumed proportional to distance, $C_n\sqrt{N}$, the assumed costs of actual shopping, and C_i, the opportunity costs of other shopping opportunities foregone. p(N) is the probable satisfaction function, and w and v are the subjective weights assigned by the consumer when he evaluates the size of each element in the equation.

The economic implications from this statement are many. For example, the minimum number of items necessary to induce a customer to shop at a given store will increase with D. Maximum shopping distance is given by the equation of the indifference curve which offers the consumer zero net benefit from shopping at a store, and is obtained by setting f(N, D) = O and solving for D to yield

$$D_n = \frac{w}{vC_d} p(N) - \frac{1}{C_d} (C_n\sqrt{N} + C_i)$$

Given the hierarchical spatial system of the theory of tertiary activity:

$p(N)$ will be a step function related to levels of goods available at each rank of the hierarchy.

$C_n\sqrt{N}$ will be a like step function.

C_i will be dependent upon the spatial characteristics of the hierarchical system.

Hence, any solution D_n will be a step function related to levels of the hierarchy. The Baumol and Ide system is entirely compatible with theory of tertiary activity. This statement is also true for the development of a total retail sales model since such a model is based upon the previous one of maximum consumer distance, and of notions of maximization of profits (which are of course based upon N). Not only do consumers discriminate among centers hierarchically and spatially, but retail varieties and sales levels and the extent of potential profits are likewise determined.

Structure of Urban Land Uses

What are the relations of the spatial arrangement of business land uses and shopping centers discussed above to the more general pattern of land uses within cities? This is the question which must next be addressed. A good picture of the complexity of patterns of urban land use has been provided by Bartholomew (1955). Bartholomew, however, described land uses without recognizing any underlying systems of order. But empirical evidence is increasing that order and systematic organization obtains in the sphere of urban land uses, and that this order is to be found in the areal functional organization of the urban area about focal points.

Perhaps one of the most pervasive types of focal point in the city is the business center. The central business district provides the most central location in the city, a point of maximum accessibility, with the highest land values. Around the central business district patterns of land use develop as accessibility diminishes and rents decrease at different rates for different types of use. In the same way, business centers below the level of the central business district provide secondary foci at other accessible points within the city. Much empirical evidence to this effect is accumulating. Both William-Olsson (1940) and Dickinson (1947) have commented upon "regionalizations" within the city which result from intraurban focal orientations. Harris and Ullman (1945) suggested a heteronucleated urban structure, and Rolph (1929) has provided information of value in this respect. Mayer (1942) and McKenzie (1933) both provided evidence of the peaking of land values at major traffic intersections, differential peakings according to the relative importance of the intersections, and the association of such peaked land values with business use of the land at and around the intersections. The conclusion is evident. Since business foci provide one observable set of central locations and focal points in present complex patterns of urban land uses and rents, they can provide clues to the eventual understanding of the larger problem of competitive bidding for urban land and resulting patterns of land uses.

Land Uses and the Rent System

How exactly does this relationship between business centers and the spatial structure of land uses work? To answer this question, it is first necessary to review the theory of urban land uses and land values.

The original outline of the theory of urban land use is today usually attributed to the work of R. M. Haig (1926, 1927) although Hurd (1911) worked on the problem earlier. Numbers of studies have been completed since those of Haig: for example, by Dorau and Hinman (1928), Weimar and Hoyt (1939), Ely and Wehrwein (1940), Ratcliff (1949), Firey (1947), and Hawley (1950). Conceptual contributions of these, however, have been limited; most have served only to restate Haig's theory. In the present section the device used is to structure the work of Haig, relating where appropriate to such supplementary studies as those of Isard and Ratcliff, etc. Thereafter, implications for patterns of business land use of existing empirical and conceptual contributions to the understanding of urban retailing are discussed.

Haig and subsequent contributors recognized that a basic ordered system of land uses resulted from operation of economic processes in society. Ratcliff (1949), for example, has written that "the locational pattern of land use in urban areas results from basic economic forces, and the arrangement of activities at strategic points on the web of transportation lines is a part of the economic mechanism of society."

Pertinent economic forces and mechanisms may be outlined as follows:

1. Each economic activity has an ability to derive utility from every piece of land. The utility is measured by the rent which the activity is willing to pay for the land.
2. The greater the derivable utility, the greater the rent which an activity is willing to pay. In the long run competitive bidding for land will be such that each site is occupied by the highest and best use. This use is the use which can derive the greatest utility from the site, and is therefore willing to pay most to occupy it.
3. As an outgrowth of the occupation of each site by the highest and best use, there results an orderly pattern of land use in which rents throughout the system are maximized.

This outline of the rent mechanism in the urban case is, of course, identical with the formulation in the agricultural case by Thünen (1826), Heady (1952), and Dunn (1955). In both, patterns of land use result from competition among land uses according to their rent-paying ability; in both, an optimal pattern of individual firms and aggregative uses and intensities results, and rents in the system are maximized.

The Simplest Spatial Model

Haig, in the case of urban land use, and Thünen, in the case of agricultural location, both maintained that the optimal spatial pattern of land use resulting from competitive bidding for land would be one in which friction of distance in the system was minimized. The dual of this is interpreted as implying that where the friction of distance is minimized, rents are maximized. Therefore, rents vary directly with accessibility. As accessibility increases, friction of distance is diminished. More funds are available to the firm to bid for the land, and rents therefore increase. Surpluses available to bid for land will be greatest for those activities which receive the greatest benefits from occupying accessible locations. Therefore, activities may be ranked according to the advantages they receive from occupying central locations. This ranking also describes ability to compete for these locations, and an orderly pattern of land use results in which sites are occupied not merely by the activity which can pay most but by the activity which receives greatest positive advantages in terms of accessibility from using the site. One ranking of this kind has been dexcribed by Ratcliff (1935, 1939).

The simplest model that illustrates this notion in the case of urban land use is the "concentric circle" scheme, which may be adapted and developed by utilizing the discussion of Isard. In this case, volume of sales (or utility in a broader sense) varies directly with distance from the city center and the effect of friction of distance, and costs of the firm vary with volume of sales. Given these circumstances, a concentric circle pattern of land use results. Rents are maximized with every site occupied by the highest and best use, and the friction of distance upon sales volumes is minimized.

Figure 3-1 illustrates in this example how utility derivable from each location by any use diminishes linearly with distance from the city center for any activity (say a retail store, x_1). This diminution is expressed by the line A-A_1. Given

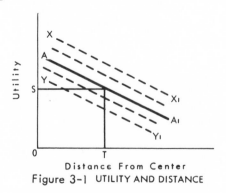

Figure 3-1 UTILITY AND DISTANCE

many retail establishments of varying efficiency, price markup, scale of advertising, etc., there will be in reality a scatter of curves about A-A$_1$, which is here assumed to be a mean for this distribution of curves. Figure 3-1 shows that at a distance OT from the city center, the dollar volume of sales (utility to the retail firm) is OS.

To achieve OS dollar volume of sales at the site will involve costs. These are accounted for in Figure 3-2. POP$_1$ is the actual returns which exceed costs ON

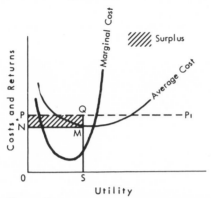

Figure 3-2 COSTS, UTILITY, AND SURPLUS

by an amount NP, and the rectangle MNPQ describes the surplus revenues (profits) accruing to the establishment. In the process of competition for the land, it is this surplus which accrues to the landowner as rent (if we assume perfect freedom of entry and competition).

Profits MNPQ are taken to accrue as rents OZ to the landowner in Figure 3-3,

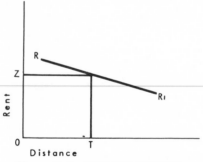

Figure 3-3   RENT AND DISTANCE

assuming perfect competition for the land. These rents OZ are, of course, those paid at a distance OT from the city center. By the procedure already outlined, rents for all other locations at all other distances from the city center may be plotted. For activity x$_1$ the resulting rent function is described by R-R$_1$.

In Figure 3-4, rent functions for three activities x$_i$, x$_j$, x$_n$ have been derived out of a set of potential uses of the land x$_1$...x$_i$...x$_j$...x$_n$. Assuming that the system has only these three uses as competitors for the land, an optimal pattern of land use has been developed. The competitive process is such that the highest and best use occupies the land. Therefore, at all distances from the city center

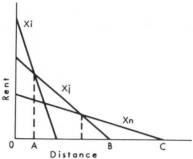

Figure 3-4   THE SIMPLEST PATTERN
OF LAND USE

to A, x_i outcompetes other activities; from A to B, x_j is the most successful competitor; from B to C, x_n is the tenant. If the system is generalized from one to two dimensions by the rotation of OC about O, then a concentric circle system of land use results.

More Complex Models

It is evident that the concentric scheme presented above is a simple generalization to facilitate understanding of the operations of lawful processes that order systems of land use. Various additional factors have been suggested to affect competitive bidding for urban land by, for example, Dorau and Hinman (1928), Wendt (1957), Weimar and Hoyt (1939), and many other students. Hoyt (1939) has also suggested an alternate model consisting of a sector rather than a concentric system, and has provided the most advanced discussion of factors which can affect residential site selection of high class residential neighborhoods. Various models of the residential sector of urban land uses are reviewed in later chapters of this volume by Marble. Harris and Ullman have suggested an additional "multiple nuclei" model of urban land uses.

Dunn (1955) commented, in the case of agricultural location, that when a variety of complicating factors is applied to the simple concentric system: "The simplicity of land use patterns is destroyed once and for all. However, this does not mean that order and system are destroyed. It means that the order imposed by the influence of economic distance takes on increasingly complex forms. . . ."

One pervasive type of more complex order is the areal functional organization of the city about focal points, one type of which is the business center. The nature of the rent system with this more complex form of order must now be clarified.

Assume the patterns of business developed theoretically above. Businesses located according to threshold requirements result in a hierarchical system of business centers as a result of the spatial operations of forces of distribution and consumption. Also, each level of the hierarchy occupies the location most accessible to its market (i.e., central to the maximum profit area at its command). Lower levels of the hierarchy serve smaller markets from sites best able to serve these markets. Lower level centers are, however, located at sites with poorer accessibility characteristics than are higher level centers. Each level of the hierarchy has particular characteristics of access to sectors of the urban market; higher levels occupy more accessible locations than lower levels.

Stratification in terms of accessibility implies stratification in terms of a general level of rent for the rank of the hierarchy; see Turvey (1957). Each level of the hierarchy has to pay a minimum rent to keep alternate potential uses from occupying the site with desired access characteristics. Since desired accessibility varies with each level of the hierarchy, it follows from the earlier discussion of the relations of utility and distance (accessibility) that so does general rent level, and the general rent comprises an additional threshold requirement of each use at each level if they are to occupy the desired site.

Some uses will pay a rent in addition to the general level. The general level is rent paid by each use at each level of the hierarchy. If any of these uses demands a special location within the shopping center, an extra rent must be paid to keep out other businesses; see Turvey (1957). This extra payment could be for location in proximity to another type of business or to a point of maximum pedestrian traffic. Within the shopping center there will be variations in rent above the general level for the particular rank of the hierarchy.

To the extent that business centers provide focal points for the system of urban land uses, it is therefore plain that hierarchically structured spatial system of rent peaks will be found within the city. To the extent that accessibility provides the key to understanding of patterns of urban land uses other than business, these land uses will be oriented to a ranked spatial system of rents, and types of uses about business foci will be directly related to the rate of diminution of rents as distance from the foci increases.

Some Additional Factors

The hierarchically structured patterns of urban land use oriented about business foci will be modified by a variety of factors, some of which are mentioned below.

Complementary and Antipathetic Uses

Spatial proximity of any pair of land uses may be so inimical to the operations of one or both that they select locations protected from each other. The most outstanding example of this in the urban area is that of the so-called "noxious" industries, which are responsible for substantial diminutions in the value of residential properties in their immediate proximity. Alternatively, locational juxtaposition may be quite beneficial, resulting in mutual increases in sales volumes, for example, in the case of retail and service business establishments where proximity reduces costs to the consumer and releases more funds for shopping.

Other Foci

Other focal points may exist in the city in addition to rent peaks associated with business land uses. Hoyt (1939) has provided the example of the homes of community leaders which act as focal points for development of high value residential districts. These and many other examples serve to illustrate that peakings in rent functions may be other than business-oriented and accessibility-conditioned.

Special Site Characteristics

Likewise, special site features may affect competitive bids for land and therefore condition patterns of urban rents and land uses in other than a business-oriented manner. Such special site features can include view property for residences, large tracts of flat land, availability of water for industry. Topographic configurations may have an impact upon the pattern of bidding for land. This impact differs according to type of land use. For any affected use the shape of the rent function is modified.

Rent and Excess Profits

In the discussion of the simplest spatial model all excess profits were taken to disappear as rent; yet, in the development of the notion of hierarchical business structure, a specific possibility of excess profits was mentioned. Also, rents have subsequently been discussed as integral parts of threshold requirements of firms. There are apparent contradictions here which must be examined in greater detail.

Rent is a payment to keep alternate uses from occupying the land, and as such it is included as part of the threshold requirements of firms. Yet it is also a payment to use the land in the face of competition between uses of similar kinds (for example, between clothing stores), and it is in this context that it was suggested that all profits must accrue to the landowner as rent. This viewpoint is entirely compatible with classical economics, within the assumption of large numbers of potential firms with perfect freedom of entry.

Now consider the formulation of the hierarchy. In this case firms compete spatially, and the most successful of each type assume their places in the hierarchical structure, paying the rent necessary to keep alternate uses from the land. This rent becomes an integral part of threshold requirements and one of the determinants of the nature of the hierarchical system. *Once established* it becomes possible for firms other than marginal hierarchical firms to compete for some surplus in their spatial context, and it becomes possible to earn excess profits.

The question remains, do these excess profits accrue as extra rent? If further firms appear and offer the landowner rent, presumably they will, for the firm offering the lowest rent will be displaced from the land. But in reality threshold requirements are likely to be of such a magnitude that conditions of entry become considerably restricted, especially for firms at higher levels of the hierarchy with larger threshold conditions of entry. Depending upon freedom of entry will be the extent to which excess profits accrue as rent, and the actual situation is likely at any one time to consist of a combination of excess profits and extra rents above needed general and specific threshold levels.

4

Empirical Verification of Concepts of Spatial Structure

IN THE PREVIOUS CHAPTER, reviews of empirical and theoretical work provided information concerning a lawful spatial ordering of business land uses. It was suggested that the business structure of the city comprises two basic conformations: (a) string street developments depending upon the importance of a street as a traffic artery, with beads of other retail uses at major intersections dependent upon centrality to residential areas; (b) business nucleations consisting of clusters of retail uses hierarchically structured from central business districts through lower ranks of clusters of neighborhood facilities. At least in part, this hypothesized structure has been developed theoretically, and the theoretical structure has been shown to be compatible with certain economic theories. Relations of the structure with the more general system of urban land uses, land values, and rents, have also been illustrated.

But several other questions remain to be asked before the structure of business may be utilized to talk about problems of highway benefits. First, it must be ascertained whether the structure may reappear in new situations using methods of analysis which develop groupings *from the data* (instead of *a priori* as did most of the previous empirical and conceptual contributions). Second, it is necessary to talk objectively on the problem of which business types succeed only when highway-oriented and which can thrive when not oriented to arterials. Third, it is necessary to understand which business types are characteristic of which level and class of business center to be able to locate exactly where highway benefits or losses will occur. It is questions of this character which this chapter seeks to answer.

Motivation

The motivation for these questions is clear. A recognizable pattern of business location habits has been noted within the framework of urban residence and pedestrian, public, and automobile transport. But the pattern of transport facilities is changing, and we need to be able to predict what the effects will be upon the pattern of business; for changes in facilities obviously will result in changes in the way people can move to satisfy their demands for goods and services and, therefore, upon the ways in which business is transacted. We need to know those kinds of business that rely most upon highway facilities and highway traffic. We need to predict sensitivity of business types and classes of business centers to changes in the availability of highway transportation. We need to be able to say *how* and *how much* the pattern of business will change as the highway system changes.

Outline of the Chapter

The chapter is empirical, i.e., the study observes business structures in reality and attempts to specify their basic ordered characteristics. First, a detailed study of the spatial groupings and locational requirements of business land uses in Spokane, Washington, is discussed. Groups of spatially associated business types are identified and these groups combined to create the basic conformations suggested by the works reviewed in Chapter 3.

Value of the empirical verification of two conformations and their constituent subgroups is revealed when findings of the Spokane study are compared with additional data collected in other areas for a range of city sizes at different time periods. These additional data comprise inventories of business land uses by location in the business centers of:

1. Cedar Rapids, Iowa, half the size of Spokane;
2. Phoenix, Arizona, of size equal to Spokane, but data only for 26 planned shopping centers;
3. Cincinnati, Ohio, twice the size of Spokane, data for two time periods;
4. a sample of planned shopping centers of various sizes across the nation.

In light of these comparisons, groupings of business types are refined considerably, and an understanding is obtained of the basic character of unplanned and planned business structures.

Refined groups are then used to study the structure of business within a 25-mile stretch of unlimited access interstate highway, U.S. 99 north from Seattle into Everett. From this study there are finally established:

1. those business types and groups which nucleate;
2. those business types which nucleate, but which when nucleated are most successful if oriented to urban arterials in string streets;
3. those business types which are highway-oriented both within and outside, or solely outside, nucleations;
4. those business types which are apparently indifferent in terms of location-- either located haphazardly or found both in nucleations and on strings with but little differentiation.

These findings, together with those for Chapter 5, are then used in Chapter 6 as the basis for a study of the actual impact of highway improvements upon business structure, an evaluation of urban planning of business land uses, and an examination of new locational opportunities created by the interstate system.

Location of the Empirical Work

The site selected for the initial and basic empirical work was the city of Spokane, Washington. Spokane has a population of 185,000 within its legal limits, and perhaps 40,000 additional people reside in its larger metropolitan area. It has long been pre-eminent as the regional capital of the area known as the "Inland Empire," serving mining and forest activities to north and east, and agricultural communities to south and west. In recent years the city has also experienced considerable expansion of facilities for manufacturing aluminum within its metropolitan area. Many detailed studies are available elsewhere which describe the history, physiography, regional importance and functions, employment and population characteristics of the city (see Berry, 1958, for an outline). These preliminary comments serve to orient the reader; major concern is with the spatial structure of those retail and service businesses which serve requirements of the population within the legal limits of the city.

Data Utilized

A complete census of business was undertaken in 1957 in the city of Spokane for each of the 296 business centers of the city outside the central business district; see Berry (1958). The census was compiled by field observation, utilization of available directories, and by checking data against records of the City Plan Commission. Since the purpose of the study was to test concepts relating to business structure external to the central business district, field work did not extend into this latter area.

The aim of the census was to count number of stores in each of 60 types of retail and service business located in the 296 business centers of the city. A business center was defined as one or more business uses spatially separated from other business uses by alternate types of land use (residential, wholesale, manufacturing industry, etc.). A problem of classification developed along major arterials leading north and east from the central business district. Here extended ribbons of business uses were beaded at intervals by shopping centers. It was observed that in the latter centers pedestrian traffic was high, in the former slight; in the latter, where businesses were surrounded by parking lots or special parking facilities, consumers made several shopping visits per shopping trip, in the former parking was frequently roadside, and customers made single-purpose visits to special stores. Also, in the former case many businesses were not visited by consumers; they served as offices and centers for functions most often performed directly in the home. Hence, it was thought appropriate to distinguish between sections of business use along major arterials. The process of delineation was based on strong intuitive notions of difference mentioned above and the degree to which these differences could be observed in the field. The classification of business types used was that of the City Plan Commission of Spokane, and is as follows:

A. Food Group
  1. Grocery, including supermarket
  2. Meat
  3. Fruit and vegetable
  4. Confection
  5. Dairy
  6. Bakery
  7. Food locker

B. Eat and Drink Group
  1. Restaurant
  2. Bar

C. General Merchandise
  1. Department and dry goods
  2. Variety

D. Apparel
1. Clothing
2. Shoe
3. Other
E. Furniture
1. Furniture
2. Appliance
3. Radio and television, sales and service
F. Automotive
1. Auto dealer
2. Used auto
3. Auto accessories
4. Other auto
G. Gas
1. Gas
H. Lumber
1. Lumber yard
2. Building supplies
3. Hardware
4. Farm equipment
I. Other Retail
1. Drug
2. Second hand
3. Feed and seed
4. Jewelry
5. Sporting, bicycle, etc.
6. Florist, nursery, etc.
7. Gift and novelty
8. Music and hobby
9. Photographer, audio-visual, etc.
10. Office equipment
11. Printing

12. All other retail
J. Office and Bank
1. Office
2. Doctor
3. Dentist
4. Real estate and insurance
5. Lawyer
6. Other professions
7. Bank
8. Post office
K. Service
1. Barber
2. Beauty
3. Cleaner and laundry
4. Funeral
5. Shoe repair
6. Miscellaneous business
7. Business services
L. Hotel, etc.
1. Hotel
2. Motel
M. Repair
1. Auto repair
2. Miscellaneous repair, including plumbing
N. Amusement
1. Theater
2. Other amusement, including billiard hall and bowling alley
O. Schools
1. Music and dance schools
P. Other
1. Mission

Certain classification problems are obvious. When, for example, does a general store become a grocery, and when does it constitute a variety store or department store? How are establishments which provide joint functions (for example, gas and auto repair, or auto dealer and used auto) to be classified? In cases such as the former, questions in the field usually served to provide the necessary answer. In cases such as the latter, two entries were usually made, one for each type of business.

Use of simple counts of business has its problems, for no account is thereby taken of the variations in size of stores, amount of business undertaken, etc. Various alternate measures were possible. Perhaps most satisfactory would have been to know total volume of sales for every type of business in each center. Unfortunately, such data were not available in sufficiently detailed amounts. A measure could have been made of total front feet occupied by each type of business in each center. This method, too, has its limitations. There are great variations in the productivity per front foot of establishments within store types as a result of the recent rapid evolution of retailing techniques. Intermixture of stores with older and newer techniques both wasteful and conserving of store area, especially between older and newer sections of the city, would seriously bias measurements. Use of total number of employees in each type of business in each of the centers as a measure was likewise possible. This measure also has many limitations, given rapid evolution of retailing and more widespread advent of self-service. Because of these limitations, and nonavailability of ideal data relating to total volume of sales by type by center, the technique used was to select the simplest of alternate counts, the number of businesses of each type.

Initial Data Processing

Data collected were then prepared for analysis. The first step was to eliminate meaningless categories such as miscellaneous business, other auto, other clothing. It was not considered fruitful to maintain these types for purposes of subsequent analysis. Also disregarded for subsequent purposes were certain types of business which had been isolated but which on examination proved to contain a great many members not performing retail or service functions for the urban population. Such categories included office, farm equipment, and confection (which included manufacturing establishments). Special types associated with local conditions were also excluded, such as business services in proximity to the central business district and second hand stores in the oldest section of the city's residential area, now degenerating to slums. Missions were left in the study for subsequent comparative purposes. Finally, also to be excluded were feed and seed stores (mostly providing the seed function) and photographic and audio-visual establishments, the argument being that these types of business were of haphazard locational pattern. For purposes of comparison, a third member of this group, the gift and novelty store, was retained.

During this process of rationalization of types of business prior to analysis a ranking system of the 60 types of business was developed based upon frequency of occurrence in the 296 centers. Frequencies for the 49 types of business subsequently analyzed are recorded in the list of groups of business types below.

Reduction of total numbers of types of business from 60 to 49 resulted in a reduction of the number of centers considered from 296 to 285. Eliminated were centers 80, 86, 161, 221, each possessing a single isolated store in one of the "other" categories; centers 35, 102, 145, 207, 223, 232, 249, each possessing an isolated feed and seed, home photographer, dance school, or farm equipment establishment.

Final analysis of the business structure of Spokane was therefore performed utilizing a census of numbers of stores of each of 49 types located in 285 business centers of the city.

The Analysis

Consider that each of the 49 remaining types of business has been mapped, and there are 49 maps of Spokane each containing information relating to numbers of stores of a particular type at 285 locations. The problem for empirical study can be thought of in these terms: What is the composite pattern of correspondence of the 49 geographical distributions? What geographical associations of business types occur? What implications do geographical clusters of business types have for the pattern and structure of business centers?

Requirements in terms of analytic techniques are two: first, provision of a measure of association of geographical distributions; second, an objective method of developing from the data tendencies for distributions to cluster and describing objectively classes of clustered variables.

Spatial Association of Business Types

The Pearson product-moment correlation coefficient was used to estimate the spatial association of each pair of business types in the 285 centers of Spokane, by correlating the number of stores of each type in the set of 285 centers.

High speed data processing machinery (I.B.M. 650 magnetic drum digital computer) was used to calculate 1,225 correlation coefficients along and above the diagonal of a symmetrical 49 x 49 matrix containing a correlation coefficient for each pair of business types. It is obviously impractical to reproduce a matrix of this size here, although Table 4-I presents one small segment of this matrix for illustrative purposes.

TABLE 4-I

EXAMPLE CORRELATION MATRIX OF FIVE BUSINESS TYPES

33	37	39	42	46	Business Type	
1.00*	.71	.60	.57	.37	33	(Department store)
	1.00	.48	.54	.43	37	(Sporting goods, bicycle)
		1.00	.49	.45	39	(Other professions)
			1.00	.56	42	(Shoe)
				1.00	46	(Bank)

*A value of $r = 1.00$ indicates perfect correlation.

Identifying Groups by Linkage Analysis

Using the 49 x 49 correlation matrix, groups of spatially associated business types were achieved using the linkage analysis technique of L. L. McQuitty (1957).

McQuitty's linkage analysis provides a technique whereby, given a matrix of correlation coefficients between a series of variables, groups of associated variables may be derived objectively from the data with no need on the part of the researcher to define limits to his groups. Both members and limits emerge naturally in the process of analysis, and the system of grouped variables is thereby derived objectively on the basis of the degree to which variables are associated (as measured by the correlation coefficient).

Since linkage analysis of this type has never previously been applied to problems of the current type, the method, assumptions, and procedures must be outlined with care.

Linkage analysis is an objective method capable of classifying variables into a number of groups or types determined by the data (a matrix of measures of association of each of the variables with every other variable). A group is defined as a category of such a nature that every variable in the category is more like some other variable in the category than it is like any variable outside the category. The method may be looked at in these terms: Suppose there are 5 maps of geographical distributions of business types. A measure of association of each map with every other map may be made. This measure may be a correlation coefficient, although it is possible to have any other measure of association. A matrix of correlation coefficients is prepared. In terms of the correlation coefficients, every variable in a type has a higher correlation coefficient with some other member of the type than with any variable not in the type. The limits to each group or type are objectively defined by the association of each variable with other variables in terms of its highest coefficient of correlation. The mathematical bases of this method follow.

Let r'_{ij} represent the index of association between any two variables such that the variable i has its highest r with variable j, and variable j has its highest r with i. This is a reciprocal relationship. Every matrix has at least one reciprocal relationship because every matrix has one or more highest r.

Assume a matrix with $n/2$ reciprocal relationships. It is obvious, then, that $o < n/2 < (m/2)+1$, where m is the number of variables in the matrix and n is the number of variables sharing in reciprocal relationships. Every n is assigned to the type represented by the pair of which it is a reciprocal member. Call all assigned variables the i_n and all unassigned variables the i_u. For every matrix all i_u have their highest r with i_n. To prove this assume the contrary. There is no i_u which has its highest r with an i_n. In this case there must be a highest r within the remaining i_u; this r would be the largest of all and therefore reciprocal, and these variables would therefore be i_n. But no i_u are i_n; therefore a fallacy exists, and the converse must be the case. Every i_u has its highest r with an i_n.

After all reciprocal pairs are identified and the cores of groups therefore established, all that is necessary is to assign the i_u to types which contain the i_n with which they have their highest r. The i_u therefore join types according as they have their highest r with members of reciprocal pairs or they have their highest r with nonreciprocal variables already assigned to a type. The method continues until all variables are assigned to types, and the limits of types therefore emerge naturally from the data.

Using McQuitty's technique and definitions of groups (that every member of a group is more like some other member of the group than it is like any member of any other group), business types were grouped into spatially associated types, and these groups are listed below from A to I in Table 4-II.

Larger Conformations

For each of the groups of correlation coefficients in the 49 x 49 matrix associated with the above groups of business types, an average correlation coefficient was calculated. To the resulting 9 x 9 matrix of correlation coefficients (each pair representing two groups of business types), linkage analysis was applied. The purpose of this was to establish whether the groups of spatially associated business types themselves had any distinctive patterns of association. Two over-all conformations of groups of business types, nucleated and arterial, emerged.

TABLE 4-II
GROUPS OF SPATIALLY ASSOCIATED BUSINESS TYPES

Group	Member of Group	Frequency of Occurrence*
A	Gas	.38
	Restaurant	.22
	Auto repair	.15
B	Grocery	.59
	Barber	.20
	Cleaner and laundry	.20
	Drug	.14
	Real estate and insurance	.13
	Hardware	.12
	Beauty	.12
	Theater	.01
C	Clothing	.08
	Variety	.07
	Dairy	.04
	Jewelry	.04
	Lawyer	.03
	Post office	.03
D	Department	.04
	Sporting goods and bicycle	.04
	Other professions	.04
	Shoe	.03
	Bank	.03
E	Bakery	.05
	Auto dealer	.05
	Used auto	.05
	Florist and nursery	.04
	Food locker	.03
	Music and hobby	.02
	Hotel	.02
F	Building supplies	.14
	Bar	.11
	Radio-TV, sales and service	.10
	Shoe repair	.09
	Furniture	.07
	Auto accessories	.07
	Appliance	.06
	Other retail	.06
	Miscellaneous repair including plumbing	.06
	Lumber yard	.05
	Gift and novelty	.03
	Motel	.02
	Mission	.01
G	Meat	.03
	Fruit, vegetable and produce	.01
H	Doctor	.05
	Dentist	.05
I	Printing	.05
	Office equipment	.02
	Funeral home	.02

*1. 0 means occurrence in all 296 centers. 0 means that type does not appear.

The *nucleated conformation* comprised three "shopping groups": B, C, and D, and the "auto row group" E. This is, of course, evidence of the validity of previous empirical work in recognizing spatial types of business.

The *arterial conformation* comprised the "auto group" A and the "supplies-repair group" F. Also within this conformation fell the two relatively minor groups G and I. This spatial group is akin to, but not identical with, the "string street" idea of previous work, and has brought out more objectively the character of this spatial type of business.

Group H, the "clinic group, " linked only weakly with the nucleated conformation and, hence, was treated as a group apart from the two conformations above. Doctor and dentist had a correlation r = 0. 77. But these types correlated with only real estate and other professions outside the clinic group at a level greater than r = 0.30.

Differential frequency habits of the groups within the conformations should be noted. Table 4-III averages the frequency figures by groups found in Table 4-II. These differential frequency habits are of great significance when the problem of classes of centers defined upon groups of business types is discussed, for dif-

TABLE 4-III
FREQUENCY OF OCCURRENCE OF GROUPS OF BUSINESS
TYPES IN SPOKANE CENTERS

	Business Type	Frequency of Occurrence
Nucleated conformation:		
Shopping group	B (with theater)	.19
	(without theater)	.21
	C	.05
	D	.04
Auto row	E	.04
Arterial conformation:		
Auto group	A	.25
Supplies-repair group	F	.07
Others	G	.02
	I	.03
Clinic group	H	.05

ferent levels of center result from differences in the frequency of occurrence of particular groups of business types.

Classes of Business Centers

With spatial groups and conformations of business types recognized, the second requirement of the empirical work in the Spokane study was to evaluate the implications of these spatial associations for the typology of business centers within the city.

Over-all Classes of Centers

The data were prepared in the following manner. Spokane possessed 142 centers with a single type of business. Of these, 84 were isolated groceries, 21 were gas stations, and another 28 were drawn from groups A, B, and F. Therefore, on the average, centers with one type of business had 0.59 groceries (84/142), 0.19 gas stations (21/142), etc. Similar averages were computed for the 44 centers with two types of business, the 14 with three, 7 with four, 15 with five, and so forth. These "average centers" were then correlated, each with every other, and a correlation matrix prepared. Linkage analysis was applied to the matrix and provided a grouping of the average centers into four classes, hereafter numbered I-IV.

Class I centers possessed from one to five business types on the average; Class II centers possessed from six through fifteen types; Class III centers were found to have from sixteen to thirty business types; and Class IV centers, the largest, from thirty-three to thirty-four. Figure 4-1 is a map of these centers.

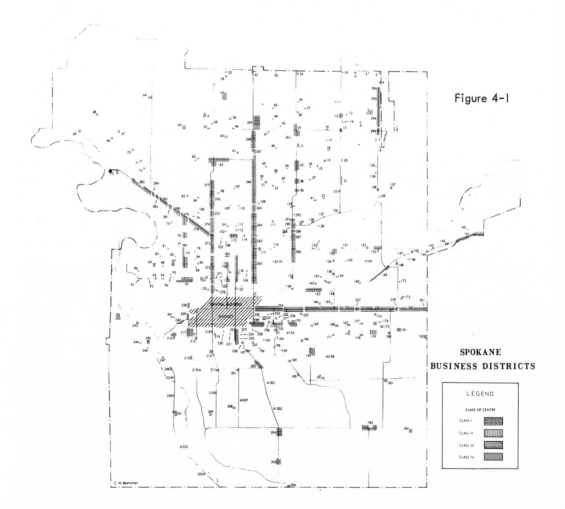

Figure 4-1

SPOKANE
BUSINESS DISTRICTS

LEGEND

CLASS OF CENTER

CLASS I
CLASS II
CLASS III
CLASS IV

C. W. Beetschen

Of the business centers of the city of Spokane, 222 fell into Class I, 45 into Class II, 16 into Class III, and 2 into Class IV. What is the functional character of these classes of centers? Table 4-IV shows the association of Class I

TABLE 4-IV

AVERAGE OCCURRENCE OF BUSINESS TYPES BY GROUP IN CLASS OF CENTER, SPOKANE

Class of Center	Group of Business*								
	B	C	D	E	H	A	F	G	I
I	1.33†	0.11	0.03	0.14	0.07	0.85	0.41	0.00	0.00
II	3.62	0.93	0.21	0.46	0.22	2.02	2.26	0.22	0.19
III	5.44	2.07	1.45	1.67	0.59	2.24	6.39	0.43	0.81
IV	7.00	5.50	5.00	2.50	2.00	2.00	8.50	0.50	0.50
Total Possible	8.00	6.00	5.00	7.00	2.00	3.00	13.00	2.00	3.00

	Expressed as a Percentage of Class									Total
I	45.2‡	3.7	1.0	4.7	2.3	28.9	13.9	0.0	0.0	100
II	35.7	9.1	2.0	4.5	2.1	19.9	22.3	2.1	1.8	100
III	25.7	9.8	6.8	7.9	2.7	10.6	30.2	2.0	3.8	100
IV	20.8	16.4	14.9	7.4	5.9	5.9	25.3	1.4	1.4	100

*See Table 4-II.

†A value of 1.33 indicates that a center of Class I has on the average one and one-third types of business drawn from Group B.

‡I.e., 45.2 per cent of the business types in a Class I center are members of Group B.

centers with types of business in groups B and A, with some drawn from group F. In Class II centers there are greater ranges of functions in the B and A groups, group F business types are as numerous as group A types, and there are some types from group C now appearing. Class III centers possess many more types of business drawn from groups B, C, and F and functions from groups D, E, and I also appear. Full ranges of business types from most groups are characteristic of Class IV centers. Particularly apparent is the contrast between the distribution of groups C and D in Class III and Class IV centers. Variation in the importance of each business group in each class of center is illustrated by the second part of Table 4-IV, which displays the percentage breakdown of each business group for each class of center. One may note, for example, the diminishing relative importance of groups B and A, the gradual expansion of groups C and F, and the very rapid rise to significance of group D, as one passes from Class I to Class IV centers.

Classes of Nucleated Centers

Using *only* data for business types of the nucleated conformation, the implications of spatially associated groups of business types for classes of centers performing necleated functions were next examined. Date were prepared in the identical manner as those prepared for examination of the over-all classes of shopping centers. "Average" nucleated shopping centers were calculated on the basis of number of types of nucleated shopping functions possessed. Each average center was correlated with every other average center, and a matrix of correlation coefficients prepared. To this matrix linkage analysis was applied, and three classes of centers were obtained, called here I_A, II_A, and III_A. I_A centers possessed from one to thirteen business types, II_A fourteen through sixteen, and III_A nineteen through twenty-one.

There were 223 centers of class I_A, 3 of II_A, and 2 of III_A. Of group I_A, only 7 centers had from nine to thirteen types of business, and 218 had from one to eight. A total of 57 centers did not have nucleated shopping functions. Implications of this classification system for spatial associations and groupings of business types are to be seen in Table 4-V.

TABLE 4-V

AVERAGE OCCURRENCE OF NUCLEATED SHOPPING
FUNCTIONS BY GROUPS IN CLASSES OF NUCLEATED
CENTERS, SPOKANE

Class of Center	Group of Business*				
	B	C	D	E	
I_A	4.24†	1.14	0.55	0.26	
II_A	7.00	4.33	2.33	1.33	
III_A	7.00	5.50	5.00	2.50	
Total Possible	8.00	6.00	5.00	7.00	
Expressed as a Percentage of Class					Total
I_A	68.4‡	18.4	8.8	4.2	100
II_A	46.6	28.8	15.5	8.8	100
III_A	35.0	27.5	25.0	12.5	100

*See Table 4-II.

†I. e., there are 4.24 Group B business types in an average Class I_A center.

‡68.4 per cent of a Class I_A center's business types are on the average members of Group B.

Class I_A centers draw most of their functions from group B, with but few from group C. Class II_A centers have a full range of group B functions, and expanded numbers of group C functions. Class III_A centers are distinguished by a full range of functions of B, C, and D. Relative changes in significance of each group of functions for the various classes of center are illustrated by the percentage figures in Table 4-V.

Classes of Arterial Centers

Implications of spatially associated groups of business types for different classes of arterial centers were evaluated in the same manner as above. Using data *only* for arterial businesses in the centers, preparation was exactly as above. "Average" centers were calculated on the basis of number of types of arterial business they possessed. A correlation matrix of each average center with every other center was constructed, and linkage analysis was used to develop four classes of centers, Classes I_B - IV_B. Class I_B centers possessed one to three business types, II_B had four to five, III_B six through eight, and IV_B nine through fifteen.

Out of the 285 centers in the city, 122 did not possess functions of this conformation. Of the rest, 116 fell into Class I_B, 19 into Class II_B, 15 into Class III_B, and 13 into Class IV_B. Implications of this structure for the geographical distribution of groups of associated arterial business types are presented in Table 4-VI.

TABLE 4-VI
AVERAGE OCCURRENCE OF ARTERIAL FUNCTIONS
BY GROUP IN ARTERIAL CENTERS

Class of Center	Group of Business*				
	A	F	G	I	
I_B	1.35†	0.72	0.00	0.01	
II_B	2.10	2.25	0.05	0.10	
III_B	2.43	4.17	0.19	0.23	
IV_B	2.52	7.67	0.55	1.26	
Total Possible	3.00	13.00	2.00	3.00	
Expressed as a Percentage of Class					Total
I_B	64.9‡	34.6	0.00	0.00	100
II_B	47.7	51.1	0.11	0.22	100
III_B	34.6	59.4	2.7	3.2	100
IV_B	21.0	63.9	4.5	10.5	100

*See Table 4-II.
†I. e., there are 1.35 Group A business types in an average Class I_B center.
‡64.9 per cent of a Class I_B center's business types are on the average members of Group A.

It is evident that the four classes are distinguished by differing mixes of group A and F businesses. Groups G and I are at all times insignificant, and only in Class IV_B centers does group I show any tendency to more widespread occurrence. Class I_B is closely associated with group A functions, with a scatter of group F. In Class II_B there is in most cases a full range of group A businesses, and more F business types are seen. Class III_B is characterized by an expanded range of F functions and the first appearances of G and I groups. Further increases of F, G, and I groups are characteristic of Class IV_B centers. The relative diminution of group A, and the rise to dominance of group F is well illustrated by the percentage figures in the lower half of Table 4-VI.

Significance of the Findings

Table 4-VII consists of a random sample of Spokane business centers and provides clear evidence of the significance resulting from possession of either nucleated or arterial business types, for business centers' functional character.

TABLE 4-VII
SAMPLE OF SPOKANE BUSINESS CENTERS

Conformation	Group	Type	62	294	255	273	254	258	268	272	269	263	204	29	79	286	163	259	69	38	241	128	26	94	250	19	23	22	3	1
Nucleated Conformation	Shopping Groups	Grocery	2	4	6	3					1	2	3	2	2	3		1	1	1	1		1	1		2			1	1
		Barber	4	5	3	2	1	2		2	2	1	2	2	1		1	1	1		1		1			1				
		Cleaner, laundry	3	4	2	4		1			3		1	2		3	1		1	1	1		1							
		Drug	2	2	2	2		1	1	1	2	1	1	1					1	1			1							
		Real estate, insurance	7	5	4	2	1	1	5		2	1	1								1									
		Hardware	3	2	3	2	1		1		1	1	1			2														
		Beauty	2	2	1	2					1	1	1	2		1								1						
		Clothing	6	1	2	1					5		1																	
		Variety	2	3	1	2					1	2		1																
		Dairy	1			1					1																			
		Jewelry	2	1	2	1					1	1		1																
		Lawyer	2	3	1					1	5	2																		
		Post office	1	1							1	1	1																	
		Department	1	8	2	1		1			2										1									
		Sporting goods, bicycle	1	3	1	2						1																		
		Other professions	1	3			1	2		2																				
		Shoe	2	2		1	2				1																			
		Bank	1	1		1																								
	Auto Row	Bakery	1	1		1	4						1	1																
		Auto dealer						8	2																					
		Used auto			2			8	8	1		1																		
		Florist, nursery	1			2	2				1		1																	
		Food locker	1					2			1																			
		Music, hobby						1					1																	
		Hotel	1					1																						
Arterial Conformation	Basic Arterial Groups	Gas	5	2	2	2	3	7	7	4	3	2		4	2		3	1	1	1		1		1	1				1	1
		Restaurant	5			3		3	5	2	1	1	2	1	1	2	2	1							1					
		Auto repair	1			1		1	2	1				1		1	2			1		1								
		Building supplies	4	1	1	1	1	2	1	2	1						1													
		Bar	3	3	4	3	1	2	1	1		3					1	2	2				1		1					
		Radio-TV, sales, service		2	3	3	1	1	1	1		1																		
		Shoe repair	1	2	2	1					1	1	1		1															
		Furniture	4	2	1	4		3			1					1		1												
		Auto accessories	1	1	4			1	1		1																			
		Appliance	3	2	3		1	2			1	1	1																	
		Other retail, fuel, etc.		1	1	1	1	3			1	1					1													
		Misc. repair, plumbing, etc.	1	1	2			1				1					1													
		Lumber yard	1								2	1																		
		Gift and novelty			1																									
		Motel																												
		Mission		3				1	1																					
	Others	Meat	1	1			1				3																			
		Fruit, vegetable, produce		1							3																			
		Printing	1	1	1			1	1	2																				
		Office equipment			2	1					2																			
		Funeral home									2																			

Source: Berry (1958).

Nucleated Types

Centers 19, 241, 69, 204, 263, 269 are examples of centers in which nucleated functions dominate. As is evident, high class nucleated centers possess groups B and C (and the highest class groups B, C, and D), lower class centers only group B functions. Group A activities are most evident among the arterial types present.

Auto Row

Center 254 is a good example of an "automobile row," characterized by great concentration of group E business types.

Arterial Centers

Centers 128, 163, 272, 268, and 258 are examples of arterial centers--centers in which arterial functions dominate. Lower order centers have group A business types (the "auto" type); higher order centers also have varying amounts of group F types (the "supplies-repair" types).

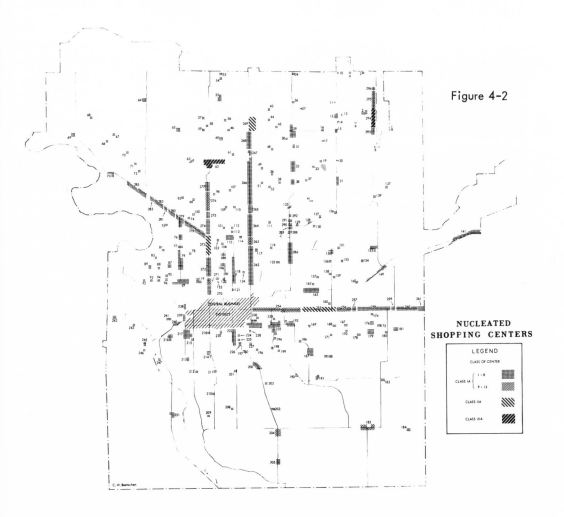

Figure 4-2

NUCLEATED
SHOPPING CENTERS

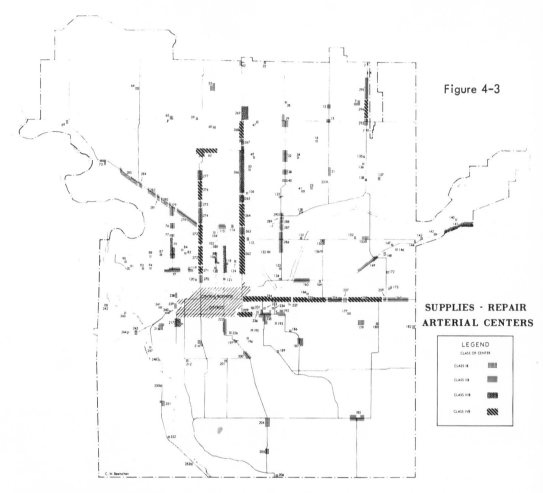

Figure 4-3

SUPPLIES · REPAIR
ARTERIAL CENTERS

LEGEND
CLASS OF CENTER

CLASS IB
CLASS IIB
CLASS IIIB
CLASS IVB

C. W. Beetschen

Joint (Complete) Centers

Centers 38, 29, 273, 255, 294, and 62 are examples of "joint centers"--centers which possess at varying levels ranges of business types of both conformations. Reliance of Class I centers upon B and A type functions is evident, and the particular significance of the grocery store and gas station within these groups shown. Alternate business types are drawn from most frequently occurring functions in groups A and B. If other types of business are to be found, they come from group F. Class II centers are defined upon more complete ranges of types of business found in groups B and A. Group F grows in significance, and a scatter of types comes from group C. Class III centers have full ranges of B and C groups, some group D functions, a full range of group A business types, and many group F. Class IV centers have full ranges of groups B, C, and D, A, and F.

Discussion

Nucleated type centers are of three levels defined on possession of group B; groups B and C; and groups B, C, and D, respectively. Arterial centers fall into

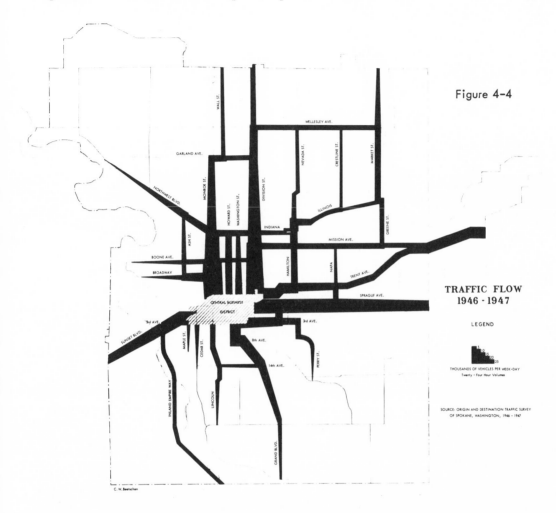

Figure 4-4

TRAFFIC FLOW
1946 - 1947

LEGEND

THOUSANDS OF VEHICLES PER WEEK-DAY
Twenty - Four Hour Volumes

SOURCE: ORIGIN AND DESTINATION TRAFFIC SURVEY
OF SPOKANE, WASHINGTON, 1946 - 1947

four levels, defined upon group A, and three stages of combination of groups A and F. Group E is of significance in defining the "automobile row" of urban centers.

Of the centers in the city 116 are completely nucleated (possessing only nucleated types). Another 57 centers are entirely arterial in character; 112 centers possess both nucleated and arterial functions, and their character depends upon the relative importance (strength of pull) of either the nucleated or the arterial business types present. But few centers can be considered "joint" or "balanced" in character.

Spatial Patterns

Figures 4-2 and 4-3 present graphic evidence of the differential location habits of the two conformations of business. Nucleated shopping centers (Figure 4-2) are widely dispersed throughout the city; arterial centers (Figure 4-3) are highly concentrated, oriented to and outlining a few major streets.

Contrasts between "nucleated" and "arterial" centers are quite evident. "Balanced" centers (for example 255, 294, 62) appear significantly on both maps.

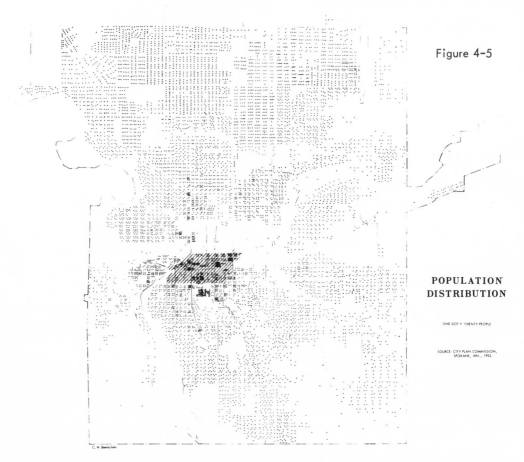

Figure 4-5

**POPULATION
DISTRIBUTION**

ONE DOT • TWENTY PEOPLE

SOURCE: CITY PLAN COMMISSION,
SPOKANE, WN., 1953.

C. W. Beetschen

Some Spatial Relations

The significance of such differential patterns is evident when these spatial distributions of the two conformations are compared with patterns of traffic flow in the city (Figure 4-4). Available data concerning traffic flow are relatively old, derived as they are from the 1946-1947 origin and destination survey for Spokane, yet obvious relationships between traffic flow and the two types of centers appear. Arterial centers are oriented to major traffic flows. Business types within this conformation seek out major axes of movement. This is suggested by the example of the traffic-oriented locational pattern of service stations and garages (Figure 4-5). There are, then, close resemblances between patterns of traffic flow and patterns of arterial centers. Business types seeking out the urban arterial have been identified.

Nucleated shopping centers are on the other hand more diffuse. However, Class III_A and Class II_A centers and the seven largest centers of Class I_A are oriented to arterials, seeking out major traffic intersections. So does center 204. Centers 256, 263, 271, and 276 are located upon the major arteries leading north and

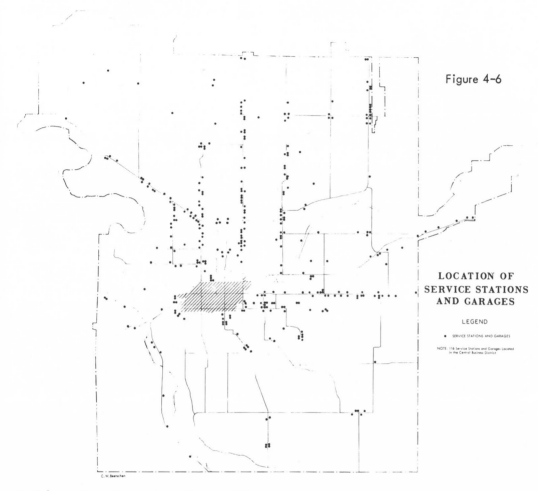

Figure 4-6

LOCATION OF
SERVICE STATIONS
AND GARAGES

LEGEND

• SERVICE STATIONS AND GARAGES

NOTE: 116 Service Stations and Garages Located in the Central Business District

east from the city center. Center 254 is an eastward extension of the automobile row of the central business district. Center 185 is small, but has expanded in the years since 1946-1947 with the extension of residences into southeastern sectors of the city. Smaller nucleated shopping centers are widely scattered, relatively evenly distributed throughout populated parts of the city, and the spatial similarities of the distribution of smaller centers and the distribution of population may be noticed (Figure 4-6).

A Summary Comment

The above empirical work has achieved the following:

1. Groups of spatially associated business types have been recognized;
2. Classes of business centers have been shown to follow from the spatial grouping habits of business types;
3. It has been possible to differentiate between groups of business types and classes of centers which are (i) nucleated, oriented to the urban population, (ii) arterial, oriented to urban arterials. Also, an "auto row" type has been distinguished.

It now remains to refine these empirical findings by comparison with data collected in other parts of the United States.

Cedar Rapids Comparison

Later chapters of the volume develop in detail a study undertaken in Cedar Rapids, Iowa, a city one-half the size of Spokane. As part of this study, using data developed for the year 1949, an analysis of the spatial association of business types was undertaken. The present note serves as a comparison of the Spokane and Cedar Rapids findings.

In Table 4-VIII, groups of spatially associated business types are identified. When linked one with another, it was found that all groups (i) - (x) linked in one conformation. Hence, only a single over-all classification of centers was undertaken. Below the central business district three classes of centers were identified: Class I (low) defined on 1-6 business types; Class II (medium) defined on 7-22 types; Class III (high) defined on 26-32 business types. There were 135 Class I centers, 25 Class II, and 3 Class III centers.

At first these results seem puzzling. Spokane groups A and F are most closely identified with a single Cedar Rapids group, (x). Group B is most closely approximated by group (i). Cedar Rapids groups (ii) - (vi) include Spokane groups B, C, D, E, and F business types. Clinic group H is group (vii), and group (viii) brings out the basis of the "auto row."

But these differences may be the result of differences in the size of the cities. From the theory presented earlier, we know that high order centers require larger thresholds for justification. Cedar Rapids is less able to justify either complex nucleated centers defined on groups C and D or complex arterial centers defined on groups F, G, and I, than is Spokane. Hence we may conclude that

1. The best defined groups in Cedar Rapids are likely to be low order [(i), basic nucleated; (v) basic service; (vii) clinic; (x) the "group A complex" of gas, restaurant and auto repair];
2. There being but little justification for high order nucleated centers, there will be less recognizable structure of high order nucleated functions [group (vi)]-- for example, the city has no department stores outside the central business district;
3. Well-defined urban arterials are less likely, and hence the single arterial group (x), reflecting lower levels of differentiation.

Also, there are differences in time period in the two studies. In eight years between the studies, variety stores, meat markets, fruit and vegetable stores have been diminishing in significance, and the supermarket has spread rapidly. It is by no means clear that today differences between groups (i) and (ii) would emerge. Group (ix) is in Cedar Rapids oriented to railroads; in Spokane it is an arterial type. This reflects differences in fuel availability and use, and the nature of procurement and distribution of lumber supplies in different parts of the country.

In summary, what appears puzzling at first is capable of some rational explanation. In so far as the business types are comparable, we can see greatest similari-

TABLE 4-VIII
GROUPS OF SPATIALLY ASSOCIATED BUSINESS TYPES
CEDAR RAPIDS, IOWA, 1949

Group	Business Types	Frequency
i	Grocery	0.67†
	Bakery	0.06*
	Drug	0.11*
	Beauty	0.16*
	Hardware	0.04*
ii	Variety	0.02*
	Supermarket	0.09†
iii	Cleaner, laundry	0.08*
	Building services, heating	0.04†
iv	Clothing and shoe	0.02†
	Other professions	0.03
v	Shoe repair	0.09*
	Fix-it	0.08
	Barber	0.18*
	Ice cream	0.04
vi	Appliance	0.04*
	Dry Goods	0.02
	Meat, fish, fruit, vegetable	0.09†
	Radio-TV, sales and service	0.07*
	Bank	0.02*
	Funeral	0.07*
	Furniture repair	0.04
	Hardware and paint	0.04
	Tavern	0.13*
vii	Doctor	0.03*
	Dentist	0.02*
viii	Used car	0.07*
	Auto dealer	0.05*
	Auto accessories	0.05*
ix	Lumber	0.04*
	Coal and fuel	0.09
x	Miscellaneous retail	0.06
	Real estate, insurance, lawyer	0.09†
	Restaurant	0.21*
	Gas	0.27*
	Auto repair	0.18*
	Farm equipment	0.07
	Theater	0.04*
	Motel, hotel	0.05†
	Other building supplies	0.06
	Furniture	0.07

*Business types which are strictly comparable with those of Spokane.

†Types where comparisons may be possible.

ties in the two situations with lower order groups supplying both nucleated and arterial functions. We expect, and observe, differences to increase as one progresses to higher order groups, for Cedar Rapids has less opportunity to justify high order nucleated centers or specialized arterials. We expect central business districts to assume, more and more, the sole responsibility for supplying high order functions as the size of the city diminishes. This tendency was also observed by Proudfoot in his studies. Nevertheless, Cedar Rapids does have both nucleated and arterial conformations, a clinic group, and the basis of an automobile row. Examination of the original data shows that Class I centers are defined on incomplete ranges of groups (i), (v), and (x); Class II centers have full ranges of groups (i) and (v) with the basic "auto" nucleus of group (x) and some group (ii) and (iii) functions. The few high order centers have a full range of all types, including group 3 (vi) and (x)--i.e., acting as "balanced" centers with both nucleated and arterial functions. Two arterials with auto rows can be identified.

Phoenix Comparison

The First National Bank of Arizona recently (1957) published data concerning business types located in 26 large- and medium-sized planned shopping centers in the city of Phoenix. These data are summarized in Table 4-IX. Interest in the data is expressed here for several reasons. Phoenix is a city of size approximately equal to that of Spokane, and comparisons may be made between medium and large business centers of Spokane (Table 4-VII) and those of Phoenix. Also, since the Phoenix data are only for planned centers, it is possible to contrast the shopping centers being evolved as a result of modern retail planning (see also Table 4-X) and the structure of centers which has evolved in Spokane not only as a result of planning but also through the play of market forces.

Comparison of Tables 4-IX and 4-VII reveals immediate contrasts. The planned centers of Phoenix are "nucleated." Neither "arterial" centers nor "auto rows" can be distinguished. This is to be expected, given the prior information from Chapter 3 (above) that planners consider arterial type developments "unhealthy" things to be suppressed. Only nucleated facilities are being developed by today's planners.

Basically, the Phoenix centers fall into two levels. The lower level is defined upon the following business types:

Group B--grocery, barber, cleaner and laundry, drug, beauty
Group C--clothing, variety
Group E--bakery.

Reference to Table 4-X will show these to be the functions most frequently found in planned centers. From the "natural" spatial associations of Spokane group B business types, real estate and hardware facilities are excluded in the lower level planned nucleations. Clothing, variety, and bakery facilities are added. Looking back to findings in Cedar Rapids (Table 4-VIII) and forward to data presented for Cincinnati (Tables 4-XI, 4-XII), it may be inferred that bakery should in fact be considered a "group B" function, rather than a "group E" type as found in Spokane.

TABLE 4-IX
PLANNED SHOPPING CENTERS IN PHOENIX

Conformation	Group	Groups and Types of Business	A	B	C	D	E	F	G	H	I	J	K	L	M	N	O	P	Q	R	S	T	U	V	W	X	Y	Z
Nucleated Conformation	Shopping Groups	Grocery	1	1	1	1	1	1	1	1	1	2		1	1	1	1	1	1	1	1	1	1		1	1	1	1
		Barber	1	1	1				1	1	1	1	1	1	1	1	1	1	1	1	1	1	1		1	1	1	1
		Cleaner, laundry	2	2	1		2	1	1	1	1	2		2		1	2	1	2	2	1	1	1	1				1
		Drug	1	1	1	1	1	1	1	1	1	1	2	1	1		1	1	1	1	1	1	1		1	1	1	1
		Real estate, insurance	1			2			1						1	1		1										
		Hardware							1	1	1		1		1						1				1			1
		Beauty	1		1				1	1		1		1		1	1	1		1		1			1	1		
		Clothing	2	8	2	8	3	1	2	1	2	2		4	1	1	1	1	1				1			1	1	
		Variety	1	1	1	1	1	1	1		1	1				1	1	1	1	1	1	1	1		1	1	1	1
		Dairy	1																									
		Jewelry	1	1		1	1	1	1	1	1	1				1					1							1
		Lawyer																										
		Post office	1		1			1			1																	
		Department	1	1		2		1										1										
		Sporting goods, bicycle				1	1						1								1							
		Other professions	2			1																						1
		Shoe	1	4	1	1	1	1	1		1										1							
		Bank	1		1			1	1	1		1	1					1										
	Auto Row	Bakery	1		1			1	1	1		1	1			1		1	1		1		1					1
		Auto dealer																										
		Used auto																										
		Florist, nursery		1		1				1				1														
		Food locker																										
		Music, hobby		1	1				1						1	1												
		Hotel																										
Arterial Conformation		Gas																										
		Restaurant	1	1	1	1		1				1													1		1	
		Auto repair																										
	Basic Arterial Groups	Building supplies						1		1	1	1	1															
		Bar					1				1																	
		Radio-TV, sales, service	1				1						3		1				1									
		Shoe repair	1	1	1		1	1							1	1												
		Furniture	1		1		1	1	1						1	1												
		Auto accessories	1								1		1															
		Appliance	1		1	1								1														
		Other retail, fuel, etc.																										
		Misc. repair, plumbing, etc.		1																								
		Lumber yard	1																									
		Gift and novelty	1	2	1				1		1				1													
		Motel																										
		Mission																										
	Others	Meat	1																									
		Fruit, vegetables, produce																										
		Printing																										
		Office equipment																										
		Funeral home																										

Source: First National Bank of Phoenix (1957).

To the lower level nucleated functions, higher level planned nucleations in Phoenix add the following functions:

TABLE 4-X

FREQUENCY OF OCCURRENCE OF BUSINESS TYPES IN 19
PLANNED SHOPPING CENTERS

Groups and Types of Business			Number of Occurrences
Nucleated Conformation	Shopping Groups	Grocery	19
		Barber	11
		Cleaner, laundry	14
		Drug	19
		Real estate, insurance	
		Hardware	7
		Beauty	14
		Clothing	14
		Variety	17
		Dairy	
		Jewelry	12
		Lawyer	
		Post office	5
		Department	7
		Sporting goods, bicycle	7
		Other professions	
		Shoe	13
		Bank	9
	Auto Row	Bakery	14
		Auto dealer	
		Used auto	
		Florist, nursery	6
		Food locker	2
		Music, hobby	4
		Hotel	
Arterial Conformation	Basic Arterial Groups	Gas	8
		Restaurant	9
		Auto repair	
		Building supplies	4
		Bar	
		Radio-TV, sales, service	7
		Shoe repair	10
		Furniture	
		Auto accessories	3
		Appliance	4
		Other retail, fuel, etc.	
		Misc. repair, plumbing, etc.	
		Lumber yard	
		Gift and novelty	13
		Motel	
		Mission	
	Others	Meat	
		Fruit, vegetable, produce	
		Printing	
		Office equipment	
		Funeral home	

Source: McKeever (1953).

TABLE 4-XI
BUSINESS CENTERS IN CINCINNATI, 1932-1933

Groups and Types of Business			1	2	3	4	5	6	7	8	9	10	11	12	13	14	15	16	17	18	19	20
Nucleated Conformation	Shopping Groups	Grocery	21	22	5	7	12	10	9	5	5	8	4	5	4	4	4	2	2	1	2	1
		Barber	9	8	1	2	3	2	1	1	2	2	1	2	1	1	1	1			1	
		Cleaner, laundry	7	3	4	2	5	3	1	2	1	1	1		1							
		Drug	7	5	2	3	2	1	1	2	1	3	2			1	2					
		Real estate, insurance	10	12		5	1	3	1	1	2		1	1	2		1					
		Hardware	4	2	1	1				1	1		1			1						
		Beauty	4	5	1	2			1	1				1	2							
		Clothing	11				1		1		1											
		Variety	12	8	1	2	1	1	1	1		1			1		1		1			
		Dairy																				
		Jewelry	2	1	1	1																
		Lawyer																				
		Post office	1	1																		
		Department																				
		Sporting goods, bicycle																				
		Other professions						1		1		1										
		Shoe		2	2					1												
		Bank	3	2	1	1	1															
	Auto Row	Bakery	4	4	1	3	1		2		1	3	1	2								
		Auto dealer	5	8							2											
		Used auto																				
		Florist, nursery	2	2																		1
		Food locker																				
		Music, hobby																				
		Hotel	1																			
Arterial Conformation	Basic Arterial Groups	Gas	13	9	2	4	4	1	1	3	4	1	1		1		1		1			
		Restaurant	8	9	1	3	2	5		1	2						1		1			1
		Auto repair	6	3	4		1		2		2	1	1		1	2				1		
		Building supplies	3		1																	
		Bar																				
		Radio-TV, sales, service	4	2	1						2											
		Shoe repair	2			2	2	3	3	1	1	2	1	1				1				
		Furniture	3																			
		Auto accessories	5	2	2						1		1	1								
		Appliance	1		1																	
		Other retail, fuel, etc.																				
		Misc. repair, plumbing, etc.	8	1	8				3	1	1		1						1			
		Lumber yard																				
		Gift and novelty	2		1																	
		Motel																				
		Mission																				
	Others	Meat	3	3	1	1		3	4			1					1		1			
		Fruit, vegetable, produce	5	3	2	2	2	1	1				1	1								
		Printing	1	1				1							1							
		Office equipment																				
		Funeral home	3	3		1		1		1			1									

*These 20 centers comprise a 20 per cent sample of the 99 centers identified in the original study.
Source: Applebaum and Schapker (1955).

Group C--jewelry
Group D--department, shoe, bank
Group E--restaurant
Group F--shoe repair, furniture, appliance, gift and novelty.

Addition of these functions of these groups is to be expected from our knowledge of upper level Spokane nucleations. These are types which require greater thresholds for their support. Prediction of exact business types is also possible using in-

TABLE 4-XII
BUSINESS CENTERS IN CINCINNATI, 1954-1955

Groups and Types of Business			Centers* (ranks: 1 2 — 4 8 5 3 — 10 9 6 7 13 — 15 14 — 11 — 12 — 19 16 17 — 18 20)
Nucleated Conformation	**Shopping Groups**	Grocery	14 15 14 6 5 7 4 2 3 11 3 5 5 1 2 4 4 1 2 1 2 1 1 1 1 1 1 2 1 1 2 1 2 1 1
		Barber	9 9 4 4 1 2 1 1 1 1 2 1 1 1 1 1 2 1 1 1 1
		Cleaner, laundry	12 5 4 4 4 7 2 1 1 4 1 1 1 1 1 2 1 1 1 1 1
		Drug	6 7 3 5 3 4 1 2 2 4 2 1 1 1 1 1 1 1 1 1
		Real estate, insurance	11 33 9 18 5 1 4 3 5 2 2 4 1 2 13 1 1 3 1
		Hardware	4 6 3 3 1 2 1 1 1 1 1 1 1 1 1 1
		Beauty	5 12 5 3 4 3 1 2 2 5 2 1 1 1 1 1 1 2
		Clothing	15 9 6 3 3 2 1 1 3 2 1
		Variety	4 5 4 1 1 2 1 1 1 1 1
		Dairy	1 1 1 1
		Jewelry	5 4 3 1 1 1
		Lawyer	3
		Post office	1 1 1 1 1
		Department	
		Sporting goods, bicycle	2 1 2 1 1
		Other professions	5 1 1 2 1 1 1 1
		Shoe	3 1 1 1 1
		Bank	3 2 1 1 1 3 1 1
	Auto Row	Bakery	4 5 3 3 2 3 1 3 1 1 1 1 1 2
		Auto dealer	6 6 2 1 1 1
		Used auto	8 3 3 3 1 3 2 1 1
		Florist, nursery	5 4 2 1 1 1 1 2 1 1
		Food locker	
		Music, hobby	1 3 1 1
		Hotel	1
Arterial Conformation	**Basic Arterial Groups**	Gas	10 13 4 5 1 4 2 4 1 1 3 1 1 1 1 1 2 2 1 1 2 4 2 1 1 2 1 1 1 1
		Restaurant	13 12 10 2 3 2 1 1 1 2 3 2 1 1 1 1 1
		Auto repair	10 2 7 2 3 4 2 2 1 1 1 1
		Building supplies	5 1 2 1 2 1 1 1
		Bar	21 16 9 7 3 1 2 1 5 1 2 1 3 2 1 2 1 1 2 1 2 1 1 1
		Radio-TV, sales, service	3 2 2 1 1 3 1 1 1 1 1 1
		Shoe repair	3 1 2 1 3 1 1 1 1 1 1 1
		Furniture	7 6 6 1 2 1 1 1 1
		Auto accessories	9 1 1 2 1 1 1
		Appliance	6 5 4 1 1 1 2 1 1 1
		Other retail, fuel, etc.	
		Misc. repair, plumbing, etc.	10 2 4 2 1 1 3 1 1 4 2 1 1
		Lumber yard	
		Gift and novelty	3 1 1 1 1 1 4 1 1
		Motel	2 2 2
		Mission	
	Others	Meat	6 5 2 3 1 1 1 1 1
		Fruit, vegetable, produce	3 1 1 2 1 1
		Printing	1 1 1
		Office equipment	
		Funeral home	2 5 2 1

*The 39 centers represent a 20 per cent sample of the 194 centers identified by the authors. Numbers of centers 1-20 refer to rank of centers identified in 1932-1933.

Source: Applebaum and Schapker (1955).

formation from Table 4-X. The exception is the addition of more furniture stores than expected.

The conclusion is, then, that planned nucleations are much like "natural" nucleations. Differences between groups B, C, and D functions again apply because all business must respond to the controls of threshold requirements. However, there are differences in business types in these various levels of nucleated centers, planned and unplanned, and these should be recognized.

Planned Shopping Centers

Some comparisons and contrasts between planned and unplanned centers have been discussed above. Attention is now turned to the planned centers. Table 4-X presents data from McKeever (1949) concerning 19 planned shopping centers throughout the United States. Applicability of these findings in the particular case of Phoenix has already been mentioned. This second set of data is presented in order that modern planned shopping centers may be contrasted with emerging trends as observed in Cincinnati, Ohio, between 1932-1933 (Table 4-XI) and 1954-1955 (Table 4-XII). Contrasts are made to answer the questions: Do planned shopping centers serve all emerging needs for retail and service facilities in the modern city? If not, what types of facilities do they *not* provide? What evidence do we have concerning the most efficient ways these nonprovided facilities are to be supplied?

It has already been shown that planned centers are of the "nucleated" type. This is readily apparent by examination of the occurrence of business types in the centers (Table 4-X):

1. Occur two-thirds of the time or more
   Group B--grocery, cleaner and laundry, drug, beauty
   Group C--clothing, variety
   Group D--shoe
   Group E--bakery
   Group F--gift and novelty
2. Occur one-third of the time or more
   Group B--barber, hardware
   Group C--department, sporting goods, bicycle, bank
   Group D--gas, restaurant
   Group F--radio-TV, shoe repair

Since the evidence is self-explanatory, little more needs to be said here, except to iterate the suggestion of two levels of planned center in Phoenix. The lower level center (neighborhood, perhaps?) is defined upon functions of set (1) above; the higher level is defined on set (2) in addition to set (1) (perhaps a community center?).

Cincinnati Comparison

Tables 4-XI and 4-XII contain data for business centers of the city of Cincinnati in 1932-1933, studied by Applebaum, and of Cincinnati and adjoining Hamilton County in 1954-1955, studied by Applebaum and Schapker (1955). These data are useful for several reasons. They facilitate further refinement of the Spokane grouping system. Evidence is provided concerning emerging trends in business. Comparisons may be made with modern planned centers in order to ascertain whether planning is filling the needs which have been emerging for goods and services.

The 1932-1933 data reveal the simpler centers of an earlier time period in a smaller city. Group B nucleated and group A arterial stores are most important. To these must be added variety stores, bakeries, shoe repair shops and miscellaneous repair facilities. Also of significance in this earlier era are group G functions: meat markets and fruit, vegetable and produce dealers. Otherwise, high order functions are highly concentrated in the three or four largest centers (centers are ranked on both tables from left to right in descending order of size).

Immediate contrasts are seen in 1954-1955. First, there are many more centers in the 20 per cent sample. This is due to (a) the difference in area studied between the two time periods; (b) growth of the system. Centers changed in relative rank, as is evidenced by the changed order in 1954-1955 of the identification numbers of centers listed in 1932-1933. Most centers grew somewhat although some, like centers 6 and 7, declined.

Variety stores and produce establishments, and also groceries, declined in numbers, but we know that growth of supermarket facilities has been responsible for this. "Automobile row" type facilities have come into existence with the pervasive automobile revolution in the United States. Standards of living have risen, and more high order facilities are to be found of the group C and D nucleated types and the group F supplies-repair type. Also, the city has grown and increasing thresholds have justified the existence of higher and higher order centers. Prohibition is a thing of the past, and bars are now to be found throughout the city.

Not only nucleated centers exist, for other than nucleated facilities have come into being. The growing United States has been characterized by growth of a variety of facilities, nucleated, auto row, and arterial. It must be evident by now that only the nucleated facilities are being provided by modern planned shopping centers. Types of business excluded from the planned centers are characteristically of another kind--those which require arterial orientation for survival and success. It is this other type of business, serving different demands, that has been labeled "unhealthy" by urban planners, who seek to obliterate it by zoning ordinances and building restrictions.

Business Locations on U. S. 99

A recent study by Shakar (1958) has presented data concerning business locations along a 25 mile stretch of U.S. 99, a major unlimited-access highway carrying the main body of traffic between Seattle and Everett (see Figures 4-7 and 4-8). Shakar's data are summarized in Table 4-XIII, cross-classified by the spatially associated groups of business types derived in Spokane and by the manner in which the particular businesses have grouped ("nucleated" [N] as in the cases of Richmond Highlands and Lynnwood, or "strip" [S]). Businesses are aggregated in strips by half-mile annules.

Differences between nucleated and arterial businesses are immediately apparent by examination of Table 4-XIII. Nucleated types stand out as discontinuous vertical tabulations of occurrences. Arterial types (especially gas, restaurant, motel) stand out as continuous horizontal occurrences along the length of the arterial from Everett (left) to Seattle (right). Accordingly, it should be possible objectively to distinguish "nucleated" and "arterial" business types with the aid of the Spokane groupings and the U.S. 99 data.

Using the U.S. 99 data, ratios of occurrence in nucleations and in strings were computed for each business type. The proportion of string occurrences was subtracted from the proportion of nucleated occurrences, and a residual index obtained. This index ranges from values of +1.0, indicating complete nucleation, to values of -1.0, indicating complete orientation to highway stretches. A value of 0.0 indicates, of course, indifference to nucleations and arterial strips (50 per

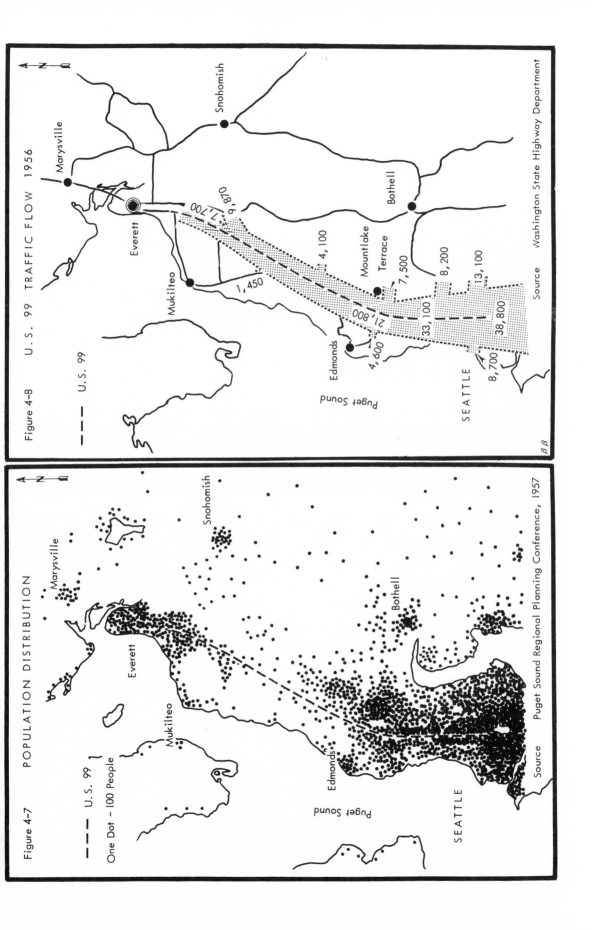

Figure 4-7   POPULATION DISTRIBUTION

– – – U.S. 99
One Dot – 100 People

Source   Puget Sound Regional Planning Conference, 1957

Marysville
Snohomish
Everett
Bothell
Mukilteo
Edmonds
Puget Sound
SEATTLE

Figure 4-8   U.S. 99   TRAFFIC FLOW 1956

– – – U.S. 99

Source   Washington State Highway Department

Marysville
Snohomish
Everett
Bothell
Mukilteo
Mountlake Terrace
Edmonds
Puget Sound
SEATTLE

9,870
17,700
4,100
7,500
8,200
1,450
21,800
33,100
13,100
4,600
38,800
8,700

TABLE 4-XIII
BUSINESS TYPES ALONG U.S. 99 BETWEEN SEATTLE AND EVERETT

Groups and Types of Business			Centers (N*) and Strips (S†)
			S N S N S S S S S S S S S S S S S S S S N S S N S S S S S S S S N S S N S N S N S
Nucleated Conformation	**Shopping Groups**	Grocery	2 2 3 1 1 2 1 2 1 1 1 1 2 1 1 1 4 1 2 1 1 1 10
		Barber	1 1 1 1 1 1 3 1 1 2 3 1 3 9
		Cleaner, laundry	1 1 1 1 2 2 1 4 5
		Drug	1 2 2 1 2 2
		Real estate, insurance	1 3 1 1 1 1 2 10 1 2 1 1 3 6 1 1 16
		Hardware	1 1 1 2 2 1 1 4
		Beauty	1 1 2 4
		Clothing		
		Variety	1 1 1 2 1
		Dairy		
		Jewelry	1 1	
		Lawyer		
		Post office	1 1 1
		Department	1 1	
		Sporting goods, bicycle	1 1 1 1 1 1 2 1 4
		Other professions	1 1 1 1 1 1 1 1 1 4
		Shoe	1	
		Bank	1 1 1 1
	Auto Row	Bakery	1 1 1 2 1
		Auto dealer	1 3
		Used auto	1 3 3 2 1 1 1 2 9
		Florist, nursery	1 1 2 1 1 1 2 1 2 1 2 1 1 1
		Food locker	1 1 1 1
		Music, hobby	1	
		Hotel		
Arterial Conformation	**Basic Arterial Groups**	Gas	2 4 4 3 1 1 1 2 3 4 1 1 2 1 1 7 3 3 2 4 1 3 5 10 2 4 4 4 1 4 1 18
		Restaurant	1 1 4 4 2 1 1 2 1 2 1 1 2 2 1 1 1 3 2 1 5 2 1 2 2 2 2 1 1 6 4 5 6 19 1
		Auto repair	1 3 2 1 1 3 1 1 1 5 1 1 1 13
		Building supplies	1 3 2 1 1 1 1 2 1 2
		Bar	1 1 1 2 1 1 1 2 1 3 1 11
		Radio-TV, sales, service	1 1 1 1 1 1 1 1 1 1 1 2
		Shoe repair	1 1 3
		Furniture	1 2 1 1 3 1 2
		Auto accessories	1 2 1 1 1 1 3 6
		Appliance	1 2 2 2
		Other retail, fuel, etc.	1 2 1 1 1 1 1 1 1 2 1 1 1 6 1 3 1 1 13
		Misc. repair, plumbing, etc.	1 1 1 1 1 1
		Lumber yard	1 1 2 1 1 1 1 1 1
		Gift and novelty	1
		Motel	1 3 3 2 3 3 2 2 1 2 2 3 1 1 4 4 4 2 2 2 1 4 1 2 6 1 3 4 1 9 2 25
		Mission		
	Others	Meat	1 2 1
		Fruit, vegetable, produce	1 1 1 1 1
		Printing	1 2
		Office equipment		
		Funeral home	1 1

*N refers to nucleated centers.
†S refers to "strips" by half-mile intervals. Columns are arranged by distance from Everett (left) to Seattle (right).
Source: Shakar (1958).

cent occurrence in each). Findings using the indices were then cross-classified with the Spokane grouping system, and these findings are recorded below in Table 4-XIV.

What do these findings tend to show? It has been possible to specify the nature of the location habits of spatially associated groups of business types. Examining Table 4-XIV, business types favoring low order nucleations (1) and high order nucleations (2) have been identified. Similarly, businesses seeking out locations within nucleations but located on urban arterials have been noted, both lower order (4) and higher order (5). It has been possible to single out uses which are always highway-oriented (3), and also to note those business types apparently indifferent in their location habits to nucleations and arterials as defined (6). These findings are, of course, of great value in their own right. But they are of even more significance when, in Chapter 6, they are actually used to evaluate the impact of highway improvements in the Marysville area, looking at data presented in the Marysville study (Chapter 5, and also Garrison and Marts, 1958) in a new light.

TABLE 4-XIV

BUSINESS TYPES CLASSIFIED BY LOCATION HABITS

1. Nucleated in Spokane, Nucleated on U.S 99

Group	Type	U.S. 99 Index
Group B	Barber	.214
	Cleaner, laundry	.552
	Drug	.454
	Real estate, insurance	.276
	Hardware	.568
	Beauty	1.000
Group C	Variety	1.000
Group E	Bakery	.549

2. Nucleated in Spokane, Less than Five Observations on U.S. 99.
   These constitute high order nucleated functions not found in the small nucleations of U.S. 99. Indices were not computed because of this limited occurrence.

Group C	Clothing
	Dairy
	Jewelry
	Lawyer
	Post office
Group D	Department
	Shoe
	Bank
Group E	Auto dealer
	Food locker
	Music, hobby
	Hotel

3. Arterial in Spokane, Arterial on U.S. 99.

Group	Type	Index
Group A	Gas	-.208
	Restaurant	-.127
Group F	Building supplies	-.116
	Radio-TV, sales and service	-.313
	Miscellaneous repair, plumbing, etc.	-.482
	Lumber yard	-.540
	Motel	-.205
Group G	Fruit, vegetable, produce	-.703

4. Arterial in Spokane, Nucleated on U.S. 99.
These are business types which seek out nucleations, but when once nucleated are found in the urban arterial situation.

Group A	Auto repair	.270
Group F	Bar	.489
	Shoe repair	1.000
	Furniture	.296
	Auto accessories	.395
	Appliance	.272
	Other retail, fuel, etc.	.126

5. Arterial in Spokane, Less than Five Observations on U.S. 99.
These are "urban arterial" business types characteristic of high order nucleations, and hence are not found on U.S. 99 in any large numbers. No indices were computed.

Group F	Gift and novelty
	Mission
Group G	Meat
Group I	Printing
	Office equipment
	Funeral home

6. Uses Indifferent to Location in Nucleations and Strings as Identified.

| Group B | Grocery | -.024 |

Groceries occur more than twice as frequently as nucleations with a cluster of Group B uses. They are therefore in a sense "proto-nucleations." But note that supermarkets increasingly favor arterial type locations with ease of auto access.

| Group D | Other professions | -.102 |
| Group E | Used auto | -.050 |

7. Uses Nucleated in Spokane, but Arterial on U.S. 99.

| Group E | Florist, nursery | -.840 |

Spokane has more florists, nucleated; U.S. 99 has more nurseries, with space-consuming uses located in the arterial "strip."

Summary

Many avenues have been explored in Chapters 3 and 4. These, both empirical and theoretical, have led to a better understanding of the structure of business and the business types associating with the several segments of this structure.

In brief, the spatial structure of business may be understood in the following terms:

1. *Nucleations.* Nucleations extend from the lowest level of the isolated grocery store, through the second level of the group B cluster, a third level includ-

ing group C uses, and a highest level adding group D. These uses are always located in urban centers (Table 4-XIV), and within urban centers in nucleated shopping centers (Table 4-II). It is readily evident both empirically and theoretically that nucleated shopping centers *within* urban areas (densely built-up areas with concentrated purchasing power) and *alternate* nucleated urban centers (nucleations in the otherwise more sparsely populated areas of less concentrated purchasing power) are of the same nature. All nucleations display significant relations to population distribution.

2. *Urban arterials.* Certain kinds of business nucleate, but when nucleated locate on urban arterials (Tables 4-XIV, 4-II). Again differing levels of performance may be noted, associated with differences in traffic flow along the highway.

3. *Automobile row.* Group E business types, Table 4-II (less bakery), are associated with the specialized automobile district of larger nucleations.

4. *Highway-oriented uses.* Table 4-XIV brings out certain business types which are successful only when highway-oriented. Within urban areas they characterize urban arterials. Between urban nucleations they are found located along well-traveled highways in "strips." Size and spacing of arterial strips are associated with amount of traffic moving along the highway; the greater the volume of traffic the greater the demands for highway-oriented business types of the gas-restaurant-motel-fruits and produce groups.

5

Impact of Highway Improvements
on Business Establishments

THE HIGHWAY IMPACT problem requires that statements be made about changes in the location structure of businesses concomitant with changes in the transportation network within which the businesses are located. Evidence of what these changes are may be gleaned from previous discussion by noting how the structure of business changes from place to place with variations in the highway system. However, the processes of site selection and competition which arrange establishments in a distinct structure take time to operate. Also, methods of merchandising, sizes of markets, and road networks within which the structure is positioned are all subject to change. Thus cross-section views, such as those presented in previous chapters, are but a glimpse of evolving and continually changing patterns. Another way to obtain evidence on impact is to observe a change in the highway system and trace out over a period of time the manner in which associated changes in business occur. Although there are difficulties inherent in this method, it is the approach in the present chapter.

Changes will occur, of course. Location at strategic places on the highway transportation network is critical to the successful operation of most business establishments. Location on the highway network influences the cost of obtaining articles for sale and distribution to customers. It influences the number of customers who can be induced to travel to the store or who will pass by in the course of ordinary travel. The location factor also bears on the price one must pay to purchase or rent facilities. This has direct implications for the ability of an establishment to occupy a site or to survive at that site.

In the present chapter indices of sensitivity are presented which are measures of the reaction of business establishments to changes in the highway transportation system--business establishments were studied prior to and following a change in the transportation system. The indices were computed using data from a study in the vicinity of Marysville, Washington, and a discussion of this study precedes the discussion of the indices. In Chapter 6 the indices of sensitivity are merged with materials on the spatial structure of business activities discussed in earlier chapters. This brings both temporal and cross-section information to bear on the impact problem.

Some Previous Studies

It is common knowledge that there are intimate relationships between the highway system and the characteristics of retail businesses. Thus, it is not surprising

100

that a number of studies were available prior to the Marysville study. Fifteen studies bearing directly on business changes as well as other studies which partially focused on business changes have been identified and reviewed by Garrison and Marts (1958a). (These studies were mentioned briefly in Chapter 1 of this volume when they were described as small area studies.) In the studies that focused directly on business changes, nearly all made by the California Division of Highways, the procedure was to survey sales during the year before a new highway facility was opened and during the year after the new facility was in operation. Comparisons were made of sales from the before to the after period.

From the standpoint of highway facilities these were by-pass studies. The particular highway facilities improved were major thoroughfares which by-passed business districts. In some cases the businesses studied were completely by-passed. In other cases the new facility paralleled the old, close by.

The studies supply only limited information on individual types of business, for they were designed chiefly to measure the well-being of whole business communities. Exceptions are cafes and gasoline stations. These were generally singled out for review and, as one would expect, they varied in sensitivity to highway change from case to case.

The study of Marysville, Washington (Garrison and Marts, 1958b), utilized essentially the same techniques as the studies just reported, but was broader in objectives. The purpose of the Marysville study was to make statements regarding the impact of highway improvements on travel patterns and many categories of land uses and land values. An attempt was made to single out a variety of kinds of retail business land uses; it is this aspect of the data that is of interest in the present study.

The Garrison and Marts study (1958b) is more general than previous studies. But the study lacks generality in that it deals with a by-pass highway which is only one of many kinds of possible highway changes. It is hoped that succeeding studies will replicate the Marysville study for other kinds of highway changes.

Summary of the Marysville Study

The Marysville study introduced above utilized the study areas of Marysville and its vicinity and, for purposes of comparison, the communities of Snohomish and Mount Vernon (Figure 5-1). Prior to October, 1954, Marysville lay astride U.S. Highway 99 a few miles north of Everett, Washington. During October, 1954, through traffic was diverted from downtown Marysville to a divided four-lane limited-access highway located just west of Marysville (Figure 5-2). In 1953 average daily traffic volume at the south entrance to Marysville was approximately 14,000 vehicles. In 1955 this average daily traffic volume was approximately 5,400 vehicles.

At the time of highway relocation, Marysville had a population of about 2,500. Its function was the service of surrounding agricultural areas with goods and services, processing of agricultural products, and other urban activities; provision of commercial services to through travelers on Highway 99; and manufacturing, especially of wood products. From the point of view of the present study interest is chiefly in retail business activities. As a shopping center Marysville proper compares

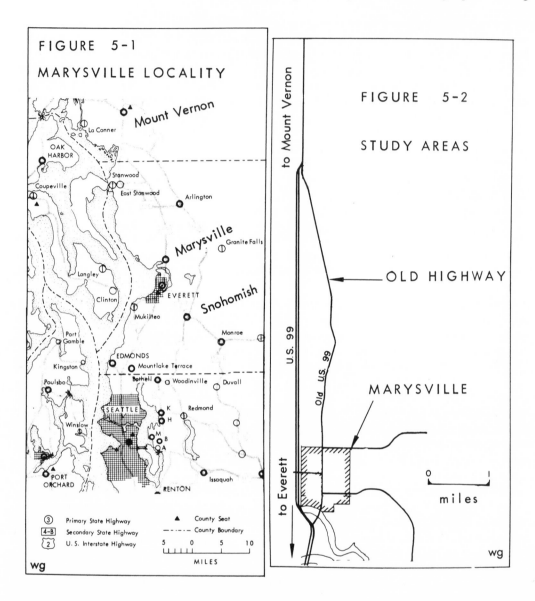

FIGURE 5-1

MARYSVILLE LOCALITY

FIGURE   5-2

STUDY AREAS

roughly with higher order (Class III "Balanced") shopping centers discussed in previous chapters, and a portion of the study area (the by-passed highway outside of Marysville) compares roughly with highway-oriented activities also discussed earlier. For further information on the nature of the study area the reader is referred to the original study and to a study of the towns of Snohomish County by Berry and Garrison (1958a). Relations between the complex of business in Marysville and vicinity and other classifications of structure are further elaborated in the next chapter of this volume.

Travel Patterns

The Marysville study was divided into three principal sections. Detailed attention was given to changes in travel patterns concomitant with the reorientation of the highway facility and reactions to the facility from the standpoint of goodness or badness. This aspect of the study was emphasized because it is recognized that the process of highway impact is carried out via changes in travel patterns.

The study used before and after interviews to determine travel patterns of persons in the rural area in the vicinity of Marysville and asked residents and businessmen in Marysville to recall travel patterns before the improvement of the highway and comment on changes since improvement. Questions were asked both in Marysville and in Mount Vernon (a community north of Marysville and also on U.S. 99; see Figure 5-1) regarding the acceptance or nonacceptance of the highway as a desired improvement.

Land Values

The second aspect of the Marysville study was a study of land value trends. Some 400 sales of property were observed in, and in the vicinity of, Marysville during the period 1953 through 1955. These were divided into the categories before highway improvement and after highway improvement and were further subdivided by types of land uses and by location relative to the new facility. Fourteen location-land use categories were recognized which, divided into before and after categories, gave a total of 28 land value trends studied.

Three location-land use categories were also studied in the community of Snohomish, located on U.S. 2 east of Everett, Washington (Figure 5-1). The land value data were converted to a 100 base and both linear and nonlinear trends were fitted. The statistical treatment allowed both visual comparisons using graphs of before and after changes concomitant with the highway improvement and the computation of a variety of indices of change.

Business Sales

The third aspect of the Marysville study was the analysis of business sales trends. This is the portion of the Marysville study of greatest pertinence to the present analysis of the impact of highways on the structure of retail business. The Marysville study used data on gross sales of business establishments in the files of the Tax Commission of the State of Washington. Records were examined for approximately 300 firms in Marysville and vicinity and another 300 firms were studied in Snohomish and vicinity, the comparison area. Records were kept for a period beginning in November and December, 1952, and continuing through July and August, 1956. The Marysville data were divided into three location categories--Marysville proper, the by-passed portion of U.S. 99 outside of Marysville, and other. The other category consisted of firms which were neither within Marysville proper nor on the by-passed highway. Data from this category are not included in the analysis within the present chapter.

The analysis in the present chapter used the Marysville observations and the by-passed U.S. 99 observations. (The by-passed highway is north of Marysville and rejoins U.S. 99 about four miles from the Marysville city limits.) In addition

Figure 5-3. Upper left: Intersection of the by-pass highway and the older highway. The view is south toward Marysville. Upper right: A view of Marysville from the by-pass highway. Lower left: The business grouping in downtown Marysville is much like that in other small or medium size towns. Lower right: Many business establishments on the old highway that formerly catered to both through and local travel are now largely oriented to local markets.

to this location classification, the business establishments were grouped into 30 business types. Not all of these would be considered retail business and consequently not all of the 30 business types appear in the present analysis. The data were reduced to a base of 100 and were subjected to graphic and statistical analysis.

An example of the analysis is furnished by the category of business, variety, and drug stores. Data were available for comparison in the Marysville area and in the control area of Snohomish. In the after period, variety and drug store sales were 89 per cent of sales in the before period while in Snohomish sales in the after period were 115 per cent of those in the before period. These measures allow the statement: the difference of 26 percentage points indicates a decline in variety and drug sales in Marysville in relation to the diversion of traffic from Marysville.

Further analysis of changes may be made from graphs of trends and associated indices. Both nonlinear and straight-line trends were fitted to the data and these are displayed for the example case in Figure 5-4. In each case the straight-line trend is downward. But, of course, this is related to the seasonality of movements. The analysis begins with a month when sales were up, relatively speaking, and ends in a month of relatively low sales. Nonetheless, a comparison of relative change is possible. In order to do this the slope of the straight line was computed. The after slope minus the before slope in Marysville yielded -2 and the same calculations yielded +9 in Snohomish. This permits the observation implied by the mean change, namely: variety and drug sales tended to decline in Marysville concomitant with the highway change.

The seasonality of sales is evidenced by the curved lines on Figure 5-4. It was desired to know whether the seasonal characteristics of sales had changed. A meth-

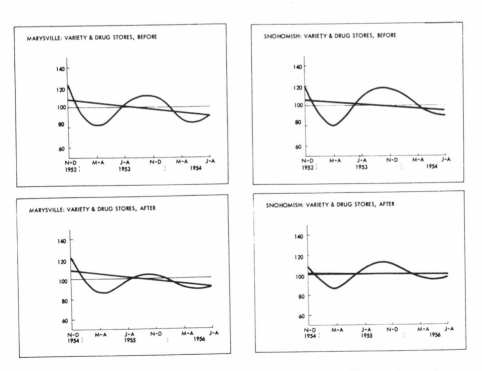

Figure 5-4. Four observations were available on the business type variety and drug stores. Both straight and nonlinear trend lines were fitted to each observation.

od was used to index seasonality, and the conclusion was reached that the change in seasonal characteristics was greater in Snohomish than in Marysville.

Computations

Data used in the Marysville study were for all firms reporting to the Washington Tax Commission for Marysville addresses. Certain of these firms are not ordinarily considered as retail business (for example, log booming) and the Marysville data were changed by deleting these activities. In addition, it was desired to restrict the present analysis to activities that occurred both in Marysville and in the Snohomish comparison area. Elimination of activities for which no comparative data were available further reduced the number of business types available for study, leaving 21 business types recognized for analysis.

Table 5-I lists the business types recognized and the number of firms observed in each type or category. Only 17 types are identified in this table, although it was mentioned that 21 types were available for study. The four types not listed in the table are not identified because of the small number of observations (firms)

TABLE 5-I

BUSINESS TYPES IDENTIFIED, MARYSVILLE AND SNOHOMISH

Business Groups and Types	Location and Number of Establishments		
	Marysville	Old Highway	Snohomish
Building material	3		5
Variety and drug	5		8
Food	9		14
General store (grocery and gas station combined)	2	2	5
Motor vehicle (new and used) and accessories	5		10
Clothing and apparel	6		7
Home furnishing and appliance	5		7
Electric appliance (radio and TV)	3		6
Hardware, paint, and garden supply	6		6
Eating and drinking place	12	10	26
Gas station	11	4	19
Garage and auto repair	7		12
Broker and law office	7		11
Hotel, motel	5		2
Cleaner and laundry	3		3
Barber and beauty	7		8
Medical services	12		15

and disclosure limitations on the tax data used in the study. However, the four types mentioned appear in ensuing tables under code numbers. It was decided to present data on these four types because the data were used when the types were further aggregated and discussed in Chapter 6 of this study.

It is unfortunate that only 21 types of business could be identified and treated. Ideally, detailed information is desired on each and every type of business. Limitations on the disclosure of the characteristics of individual firms when tax data are used prevented this fine breakdown and the classification adopted is a compromise forced upon the study by the necessity of aggregation to avoid disclosure. It is admitted that classifications of this type are somewhat subjective and may disguise conclusions that should be made or force conclusions that should not be made. Business establishments which react in exactly opposite ways to a highway improvement might be forced into a classification, for example, and yield a business type which was essentially unchanged by highway improvement.

It is hoped that when the Marysville study is replicated elsewhere the observations will be great enough in number to avoid problems of disclosure. If grouping is necessary, members of a group should be known to be homogeneous as regards highway impact. It was not felt by the authors of the Marysville study that the classifications introduced serious error into the data. It would be desirable, however, to have empirical evidence bearing on this point.

Indices Used

A good index of sensitivity to highway change would be a number which displays at a glance the character of change in business activities when the highway is changed. In the Marysville study it was desired to keep the computations of indices of change on an extremely simple level so the computations could easily be replicated and understood. In the present study the objective is to produce an index of sensitivity without regard to difficulty of computation. For this reason the method of computation here is slightly different from that used in the Marysville study.

In the Marysville study changes were measured by comparing mean sales before highway improvement with mean sales after highway improvement, comparing the slopes of linear business trends in the before period with those in the after period, and comparing the shape of nonlinear curves. The present study is restricted to comparisons of means and comparisons of the character of nonlinear trends. The comparison of linear trends was not undertaken because the information in the linear trends was contained within the information on means and nonlinear trends.

The symbols defined below are used in the description of the index number formulas:

$\overline{M}_b$ = average sales of a business type in Marysville during the before highway reorientation study period

$\overline{M}_a$ = average sales of a business type in Marysville during the after highway reorientation study period

$\overline{S}_b$ = average sales of a business type in Snohomish during the before study period

$\overline{S}_a$ = average sales of a business type in Snohomish during the after study period

M^* = value of $\sqrt{\sum_i (B_i - B'_i)^2}$ where the B_i are the parameter estimates of fourth

order polynomials ($i = 1$ to 4) fitted to business trend data for Marysville. B refers to the before period, B' refers to the after period.
S* = value for Snohomish defined as M* above

Four indices are used and these are defined below.

$$x = (\overline{M}_a / \overline{M}_b) / (\overline{S}_a / \overline{S}_b)$$

x is a percentage obtained by dividing percentage changes of means in Marysville and in the Snohomish control area. A value of 100 indicates no difference in the percentage change in Marysville from the percentage change in Snohomish; less than 100 indicates a loss in Marysville relative to Snohomish; a number greater than 100 indicates a gain in Marysville relative to Snohomish.

$$y = (\overline{M}_a / \overline{M}_b - \overline{S}_a / \overline{S}_b) / (\overline{M}_a / \overline{M}_b + \overline{S}_a / \overline{S}_b)$$

y is also computed using the information on means. A negative value of y indicates that Marysville has lost relative to Snohomish and a positive value indicates that Marysville has gained relative to Snohomish.

$$p = M* / S*$$

p uses the distance measure computed to characterize the magnitude of change in nonlinear trends. A value of 100 would indicate no difference between the magnitude of change in Marysville and the magnitude of change in Snohomish. Magnitude, of course, refers to the distance between points (see the definitions of S* and M* above). A number greater than 100 would indicate a greater change in magnitude in Marysville than in Snohomish and a number less than 100 would indicate a greater change in Snohomish than in Marysville.

$$q = (M* - S*) / (M* + S*)$$

Values of q range between +1 and -1. Like p, q is a measure using the characteristics of the nonlinear trends. Positive values indicate greater change in Marysville than in Snohomish and negative values indicate greater change in Snohomish than in Marysville.

Indices p and q relate to the character of business trends. Both indices are based on a distance measure in four space defined by the parameter estimates of the fourth order polynomials fitted to the nonlinear data. This distance measure is not a conventional one, although there are distance measures with known statistical properties, so a brief description is appropriate. Comparison problems in treating the data were too complex to allow a simple visual comparison of trends. There were four cases to be compared for each business type, two in Marysville and two in Snohomish. The distance measure was adopted as a straightforward and easily visualized method of comparing the magnitude of the polynomial regression coefficients, and the operation is defined operationally and algebraically. Characteristics of this measure are unknown so far as its use in a probability manner is concerned, but there was no attempt to measure differences from the standpoint of probability. The measure was adopted as a measure easily computed and relatively straightforward in meaning.

Sensitivity of Businesses to Highway Reorientation

Reorientation of U.S. 99 west of Marysville resulted in a decrease of about two-thirds in through traffic in Marysville. Business changes concomitant with the highway reorientation are those related to a decrease in the number of vehicles passing the business establishments. But this is an oversimplification of the effect of the highway change since the highway improvement led to several traffic and other changes. Evidence from the Marysville study either verified or gave some evidence that the highway reorientation led to the following changes which in turn influenced business sales.

1. It is clear that there was a marked reduction in the number of through travelers in Marysville. The amount of north-south traffic on U.S. 99 is greatest in the summer and least in the winter. Thus the relative reduction was greater in the summer than in the winter.
2. The highway improvement decreased traffic congestion south of Marysville between Marysville and Everett. The improvement extended from Everett north past Marysville to a point about four miles north of Marysville. Access to Everett from Marysville was simplified as was access to Marysville from Everett. There is some evidence that persons in Marysville shopped in Everett more frequently following the highway improvement than before. There is also some evidence that Marysville has taken on more of a suburban character relative to Everett with new home construction in an around Marysville related to the greater ease of access to Everett.
3. Removal of through traffic from downtown Marysville decreased congestion and improved ease of finding parking. In these ways, it increased the attractiveness of downtown Marysville as a place to shop.
4. Improvement of north-south highway facilities has changed the competitive position of Marysville and Everett versus other communities to the north, especially Mount Vernon, Washington. It is not known to what extent this is true and to what degree this selectivity affects Everett or Marysville (or perhaps Seattle further to the south) and it is not known whether this is a temporary shift or not. Plans have been made to improve U.S. 99 through all of western Washington. The area north of Everett and in the vicinity of Marysville was one of the first improvements in this part of western Washington. Construction is going on now on U.S. 99 between the present by-pass highway and Mount Vernon, and it is not known how the competitive situation of communities will stabilize when construction is completed.
5. There is evidence that the construction of the by-pass highway has made Marysville a somewhat more desirable place to live from the standpoint of residential amenities. It is presumed that this is largely related to the removal of traffic from downtown Marysville streets. This may also be related to the greater ease of access to Everett and other places because of the new highway facility. This should affect the competitive position of Marysville as a site for residential building and consequently the available market for businesses located in Marysville.

6. There is evidence that a number of business operators have changed methods of merchandising or lines of goods or services offered and changed relationships with customers as a consequence of highway reorientation.
7. It is speculated that changes have occurred in service business relationships. For one thing, Marysville is somewhat more accessible to the north than previously, and this may have bettered the competitive position of service activities. On the other hand, service activities in Everett are now more accessible to the Marysville market. This may have increased the amount of travel by persons making service trips from Everett to Marysville.

It was emphasized that there was some evidence for the points listed above. The problem is not so much whether or not they are true but the verification of the extent to which they are true. Resources were not available at the time of the Marysville study to make a complete survey of travel patterns and purchasing habits related to the listed characteristics, but it is hoped that in ensuing studies a great deal of attention will be given to codifying the travel parameters of highway reorientation. Later chapters of the present volume report basic research related to these points.

Some business had on the average higher volumes of sales prior to highway improvement than after and some had lower. Indices of sensitivity which use changes in level of business are displayed in Table 5-II. It should be remembered that business movements in Marysville were adjusted by the level of movements in Snohomish which should remove characteristics of general business trends from the variability in the data. These are business movements concomitant to the highway reorientation and independent of general business trends.

It is necessary to ask what sort of order may be found in the data. The simplest thing to do is to group movements as regards their magnitude, and this is done in Table 5-II. Inspection of index x was made with the rule that a type of business was the member of a group if its index was closer to the index of another member of that group than to the member of some other group. Electrical appliance dealers, for example, were grouped with business type medical services and business type cleaners and laundries because the index x for electrical appliance dealers was closer to that for cleaners and laundries than it was to that for business type clothing and apparel. This grouping takes into account the direction of movement. A ranking on magnitude of movements without regard to direction would have given an entirely different grouping. In Marysville, for example, garages and auto repair shops, medical services, hotels and motels, and general stores would have all fallen into the same group. This point is stressed in order to emphasize the directional characters of the indices.

There is little more that may be said from this strictly empirical point of view. It may be noted that for both Marysville and the old highway more business moved downward than upward and in each case the greatest movement of business types was in downward direction. These observations are by no means grounds for value statements regarding whether the highway improvement was good or bad for business in Marysville. In order to make such statements it would be necessary to attach value judgments to the desirability of business change of various types. No basis is available for doing this.

TABLE 5-II
INDICES OF CHANGE IN LEVEL OF BUSINESS

Business Type	Index x	Index y
Marysville		
Medical services	157.9	+ .225
Cleaner and laundry	137.9	+ .159
Electric appliance	136.5	+ .128
Clothing and apparel	125.2	+ .112
12	122.3	+ .100
Food	121.8	+ .098
Barber and beauty	114.5	+ .068
Hardware, paint, garden supply	105.3	+ .026
Gas station	102.4	+ .012
Building material	91.9	- .042
03	91.8	- .043
02	88.6	- .061
Eating and drinking place	87.5	- .066
Broker and law office	87.0	- .070
Motor vehicle sales	81.0	- .107
Variety and drug	77.3	- .128
Home furnishing and appliance	72.4	- .160
General store	66.2	- .205
Hotel, motel	64.6	- .215
Garage and auto repair	52.8	- .309
Old Highway		
05	117.4	+ .080
Gas station	91.8	- .043
Eating and drinking place	60.9	- .243
General store	53.8	- .300

It should also be emphasized that these are measurements in the short run. No data are available for a long period of time, and there is no basis for assuming that the timing of the impact is the same for each type of business. Data are needed over a long period of time to determine what ultimate changes in level take place.

The above paragraphs were written from the empirical point of view. A more important way to look at the data is in terms of the processes that are presumed to influence business change. Earlier, seven changes were listed reflecting travel and amenity impacts of the new highway facility, and it is desired to link the indices of sensitivity to these changes. This can be done only in a very gross way because of the lack of really fine scale information on travel. The impression is that those businesses which have tended upward are those (1) at which persons find it more convenient to shop in Marysville because of decreased congestion, (2) which are related to expanded residential development in and around Marysville, and/or (3) which are gaining customers because of more travel into Marysville from the tributary area to the north.

As to the negative changes, it seems reasonable to say that these are cases where business establishments were affected by the absence of through traffic and/or by the ease of traveling elsewhere to purchase goods and/or of procuring services elsewhere. The difficulty in making statements is clear. The business change effect is a result of multiple causes. What is needed is a research design that specifically takes these several causes into account (and perhaps others) and allocates them to particular types of business. No such study is available in the published literature, but it is hoped that one result of the Marysville study and the present volume will be to point up the need for this type of analysis.

Another and more meaningful interpretation of the data is from the standpoint of the spatial structure of business identified theoretically and empirically in previous chapters of this volume. Interpretation of the data in these terms is postponed until Chapter 6 of this volume.

Indices p and q relate to the characteristics of the business trends, specifically to the character of nonlinear trends. On reflection, it is easy to see that characteristics of business may be greatly influenced by a highway improvement although the level of business identified the study of means is essentially unchanged.

Index p shows whether the magnitude of change was greater or less than that in the Snohomish control area. Index q gives the same information.

Groupings in Table 5-III are with regard to magnitude of change regardless of whether or not change tended to be dampened or increased by highway reorientation.

The character of the index used is such that presence or absence of dampening is not clearly shown. The index is merely a gross measure of magnitude of change in nonlinear characteristics.

Two suggestions may be made as theoretical explanations for changes in nonlinear trends. It is known that through traffic on U.S. 99 is very seasonal. This is true because of seasonal characteristics of the tourist industry and aspects of the trucking industry related to logging and agriculture. The removal of this through traffic from Marysville would tend to alter the seasonal characteristics of businesses which cater to that through traffic. It is known that businessmen have changed emphasis on lines of goods carried since the highway improvement was made. A second suggestion to explain variation in seasonal characteristics of business is that emphasis is now on goods (or services) with different seasonal characteristics than goods previously marketed. The extent to which this is true is unknown.

Here is another case where business changes may be related to travel processes and other processes having to do with retail business, but a quantitive linkage is not possible because of a lack of basic data. Again, this emphasizes the need for research which takes into account the several processes conditioning business changes related to travel changes.

Interpretation of the indices of changes in characteristics from the point of view of the spatial structure of business is postponed until the following chapter of this volume.

TABLE 5-III
INDICES OF CHANGE IN CHARACTERISTICS
OF BUSINESS

Business Type	Index p	Index q
Marysville		
Home furnishing and appliance	3, 777. 4	+ . 948
Gas station	2, 592. 4	+ . 924
Broker and law office	1, 713. 7	+ . 895
Electric appliance	852. 7	+ . 890
12	424. 6	+ . 809
Clothing and apparel	354. 4	+ . 713
Medical services	228. 7	+ . 586
Cleaner and laundry	174. 4	+ . 560
02	168. 2	+ . 391
Eating and drinking place	149. 1	+ . 271
Garage and auto repair	141. 7	+ . 254
General store	101. 8	+ . 197
Hardware, paint, garden supply	89. 2	+ . 009
Variety and drug	67. 0	- . 057
Building material	44. 7	- . 198
Hotel, motel	35. 6	- . 382
Barber and beauty	24. 2	- . 475
Motor vehicle sales	18. 3	- . 611
Food	16. 7	- . 699
03	8. 6	- . 841
Old Highway		
Gas station	1, 650. 3	+ . 886
General store	419. 3	+ . 615
05	69. 2	- . 182
Eating and drinking place	37. 0	- . 460

Summary

This chapter has been used to review work on the sensitivity of business to high-way reorientation. On the basis of data from the vicinity of Marysville, Washington, indices of sensitivity to business change were presented. These indices of sensitivity are chiefly for the purpose of identifying expected changes in the spatial structure of retail business and, as just mentioned, interpretation in terms of spatial structure is postponed until the following chapter of the volume. Throughout the chapter comments were made on the design of research for purposes of identifying highway impact in cases such as the Marysville case. It has been emphasized that a complex set of travel and amenity changes operates concomitantly with highway change. Future research designs should incorporate methods of clearly identifying these changes. Changes in retail business depend on changes in this set of complex factors, so research designs should include devices for relating effects to several causes (such as multiple regression). Also, the period of study should be long enough for the clear establishment of trends.

6

The Relations of Highway Impact,
Spatial Structure, and Urban Planning

PREVIOUS CHAPTERS of this volume have presented findings concerning the spatial structure of business centers (Chapters 3 and 4), and have used a variety of sensitivity indices to characterize the short-run sales records of business establishments from a period before to just after the reorientation of a highway in a small urban center (Chapter 5). The present chapter seeks to integrate these approaches. In particular it attempts to associate short-run fortunes of businesses with their spatial structure. This association is accomplished in the first part of the chapter, and important tendencies both for centralization of the structure and for sensitivity to traffic volumes are noted.

If these are observable short-run trends, what of the larger period? What patterns of relocation are appearing as a result of highway improvement over longer periods of time? These questions are discussed in the second and in the final sections of the chapter.

Since relocation can only be considered within today's institutional framework of planning, the second section considers in light of previous empirical evidence abilities of the system to relocate under situations of highway improvement. Planning today generally implies urban planning, and hence an urban example, again Spokane, is used as a case study. It should be remarked, however, that planning in Spokane is much like planning elsewhere. Concepts and ideals, plans, methods, and actions are very similar.

In light of the empirical evidence, what are location tendencies induced by and location opportunities provided by highway improvements and particularly by the new system of Federal highways? These questions are reviewed in the third and final part of the chapter. First, arterial business is discussed, and new empirical evidence is brought to bear on questions of trends and tendencies with improvement of, and upon demands for, facilities. Second, nucleated businesses are considered in light of the rapid expansion of planned shopping centers since the Second World War.

Marysville Data Grouped in the Spatial System

Evidence has been presented in Chapter 5 concerning changes in average business fortunes both before and after reorientation of U.S. 99 in the vicinity of Marysville, and also relating to characteristics of business trends before and after reorientation. It was noted that an understanding of relations of these short-run trends

115

and the spatial system of business was essential to provide more conclusive meaning to observed tendencies. Hence, the Marysville data were recombined on the
basis of groups of spatially associated business types revealed in Spokane (Table
4-II) and also more refined groups based upon subsequent comparisons of Spokane
findings and U.S. 99 data (Table 4-XIV). Recombination reveals how various spatial sectors of Marysville retail and service business, nucleated and arterial, fared
in the short run under the impact of highway reorientation.

One problem of recombination is to be noted. At least five business establishments were required per spatial group before it was considered that trends of any
significance for the group could be recorded. But Marysville is an urban center of
rank less than Everett and Seattle (see Berry and Garrison, 1958a) or approximately equal to a balanced Class III center *within* Spokane. Consequently, no group sensitivity indices could be computed for the high order groups D, G, and I of the Spokane structure. In the second instance, that of the refined comparative groups, indices could only be calculated for the always nucleated and always arterial (highway-oriented) groups (Table 4-XIV).

Comparisons with Spokane Groups

Changes in level and characteristics of Marysville businesses before and after
reorientation of U.S. 99 (compared with the Snohomish control area) are available for Spokane nucleated groups B and C, auto row group E, and arterial groups
A and F. Indices presented below in Table 6-I are averages for the business establishments which were members of these groups.

TABLE 6-I

CHANGES IN LEVEL AND CHARACTERISTICS OF
MARYSVILLE AND SNOHOMISH BUSINESS ARRANGED
ACCORDING TO SPATIALLY ASSOCIATED GROUPS

Group	Indices of Changes in Level	
	x	y
Nucleated B	121.1%	+ .096
Nucleated C	82.9%	- .094
Auto row E	80.2%	- .110
Arterial A	93.5%	- .034
Arterial F	101.1%	+ .006

Group	Indices of Changes in Characteristics	
	p	q
Nucleated B	272.4%	+ .463
Nucleated C	81.3%	- .103
Auto row E	43.8%	- .391
Arterial A	192.4%	+ .316
Arterial F	139.3%	+ .164

Meanings of the indices of sensitivity x, y, p, and q have already been developed in Chapter 5. What does Table 6-I tell us? First, that the business groups most sensitive to highway reorientation and the removal of through traffic from Marysville were the nucleated groups B (grocery, etc.) and C (variety, clothing, etc.), auto row group E, and arterial group A (gas, etc.). Arterial group F, the supplies-repair group, changed but little. Groups C and E lost more business in this particular situation of highway impact, in comparison with the Snohomish control area, and group B gained most. While Snohomish was declining absolute gains were recorded by group B, absolute losses by group C. For groups E, A, and F both Marysville and Snohomish were losing business, but at different rates.

Comparisons with Refined Groups

Indices for the always nucleated and always arterial groups of Table 4-XIV are presented in Table 6-II. Codes for indices are the same as those used in Table 6-I.

TABLE 6-II

CHANGES IN LEVEL AND CHARACTERISTICS OF MARYSVILLE AND
SNOHOMISH BUSINESSES ARRANGED BY ALWAYS NUCLEATED
AND ALWAYS ARTERIAL GROUPS OF BUSINESS TYPES

Group	Indices of Changes in Level	
	x	y
Always nucleated	109.1%	+ 0.044
Always arterial	104.6%	+ 0.022
Group	Indices of Changes in Characteristics	
	p	q
Always nucleated	110.9%	+ 0.052
Always arterial	74.3%	- 0.148

These findings confirm the previous observations. Lower order nucleated uses are more sensitive than highway-oriented uses, and in the particular case of highway impact noted here, local trade has increased. At first glance, findings showing increases in level for the highway-oriented group are somewhat puzzling. It has previously been noted that group A uses declined while highway-oriented repair facilities (building supplies, and so forth, group F) improved within Marysville through the time period of traffic diversion (see also the indices presented in Chapter 5). The always nucleated group pools opposing tendencies, and on balance the trend is upward (for the items pooled see Table 4-XIV).

Highway Impact on Business Structure

The tables just presented give information consistent with that on individual business establishments discussed in Chapter 5. But here the information is in terms of the spatial structure of groups of business establishments. The evidence seems strong enough to warrant the conclusion that highway impact means a structural sensitivity leading to greater centralization of nucleated business establishments.

With greater ease of movement, purchase of high order goods in specialized centers is substituted for purchase of these goods in lower order centers. The impact of highway improvement is in terms of the more efficient operation of commercial centers.

A second conclusion from the examination of the data is that a rearrangement of arterial groups is taking place. These business establishments are sensitive to traffic volumes, and it is presumed that when they relocate in areas with higher traffic volume the result is a more efficient operation of these businesses. The latter statement is a presumption; data are not available to substantiate the point.

Centralization Tendencies

The notion of centralization is by no means a new one. In Chapter 1 of this study the notion of centralizing effects of transportation improvements was introduced from the standpoint of common knowledge and Isard's (1956) theoretical discussions. Mitchell (1939), Scotton (1953), and Chittick (1955) present evidence for this point by tracing associations of centralization and improvement of highway facilities over longer periods of time. The topic again appears in Section V of this volume with studies of the relations of highway impact and changes in medical service areas.

Data from the Marysville study show that centralization is selective and hierarchical. The nucleated group B increased the amount of business in Marysville while the amount of business of nucleated group C decreased. This variability of change was emphasized in Chapter 5 of this volume and also in the original study by Garrison and Marts (1958b) in which it was reported that some businessmen reported an increase in the amount of business while others reported a decrease in the amount of business. Business seems to have increased in Marysville in the case where the business group in Marysville has increased its competitive position versus other centers owing to improved transportation connections. Decrease in congestion within Marysville and improved availability of parking space have also undoubtedly contributed to this bettering of the competitive position of these group B types of business associated with local trade. These are also aspects of highway change.

The relative decline of the nucleated group C reflects centralization of these activities incident upon highway improvements and greater ease of access to higher order centers such as Everett and Seattle. The auto row group E has also shown great sensitivity to the centralizing effects of highway improvement. Presumably, in these cases the pay-off is to Everett and Seattle vis-à-vis Marysville, which has lost competitively because of transport improvements.

In summary, it is noted that all of the nucleated groups show marked sensitivity to centralization of facilities. In two cases (groups C and E) this led to a decrease in the amount of business owing to improved competitive position of larger centers, but in one case (group B) it led to an improvement of business locally because of a centralization into the Marysville area.

The Arterial Groups

One of the arterial groups was somewhat sensitive to highway change while the other was not. Arterial group F, the supplies-repair group, changed but little; this may be either because this group is relatively insensitive to location effects or sensitive to location effects but slow to change when the highway situation changes (at least slower than the short-run observations of change in Marysville). Whether one of these two cases is true or whether some other cause is appropriate is not known. So far as the highway impact on business volume is concerned, however, the conclusion may be made that in the short run these activities are not affected by highway improvements.

Arterial group A showed a decrease in level of business and a change in characteristics of business greater than that for arterial group F. Arterial group A contained activities which are well known to be dependent on the volume of traffic, and the decrease in volume of traffic has meant a decrease in the amount of busines. Other establishments better located as regards the highway system are obtaining the business lost by this group. This, too, is a short-run observation. At the time the data were observed, there had not been a marked relocation of businesses of this type in the vicinity of the increased traffic volumes elsewhere, but it may be that the change will be more marked over a longer period of time as response in the form of physical relocation of arterial business facilities has time to operate.

For that matter the remarks made earlier about nucleated groups are also applicable to the short run, and in this case changes may be more marked when new facilities are constructed which better reflect the availability of transportation. The question of how much change eventually occurs and how fast it occurs is as yet unanswered. Information on rates of reinvestment in response to new business opportunities is simply unavailable for many kinds of retail business. Also, whether or not reinvestment may take place depends on the amount of land available and institutional restrictions (such as zoning) on the use of that land. The degree to which zoning is flexible enough to allow business relocations consistent with opportunities created by transportation improvement exercises a critical control on the impact problem. Topics on this point are investigated in the next section of this chapter.

Planning and Highway Impact on Business Structure

Improvement of a highway system not only will occasion short-run variations in the fortunes of business establishments; it will also occasion a tendency for relocation of retail and service business. The relocated pattern should be more efficient than the previous pattern. But relocation possibilities are set within an institutional framework consisting of planning policies and practices, and it is pertinent to ask to what extent these policies and practices are compatible with efficient long-run relocation in the new situations.

The discussion to follow is restricted to urban planning because today, with but few exceptions, planning is urban planning. It is presumed that the objective of planning is to guide urban growth and redevelopment in an orderly manner which maximizes social welfare. The ensuing discussion is not intended as a blanket criticism of planning policies and practices, although several fundamental weaknesses of these in the case of urban business are noted. The question is asked: to what extent will present planning and policies *allow* pay-off from highway impact in terms of changed and relocated business relationships? To the extent that present planning policy and implementation does not allow for relocation under the impact of improved highway facilities, one may make conclusions regarding restrictions on the size of the impact and, if planning policies are to be altered, how they might be altered. The Spokane study area is used as a case in point of this inquiry.

Planning in Spokane

Planning in Spokane has historical roots extending over a quarter of a century from the implementation of the city's first zoning ordinance in 1929. A major segment of this planning for urban land uses has concerned the locations of retail and service business, and it is this segment which is of concern here. It will be noted that until recently adequate amounts of land have been zoned for business with a zoning code so flexible and lenient that patterning of the spatial structure of business in accordance with the free play of market forces has not been unduly impeded. Exceptions to this statement include bars, cocktail lounges, auto wreckers (and the like), for which more vigorous zoning controls have substantially restricted freedom of choice of location. But a new and more restrictive zoning ordinance has been proposed to serve the changing needs of the city. This new ordinance is more restrictive of land zoned, and is designed to guide the locational choice of business firms in accordance with planners' concepts of a "good" business structure (this question has been examined in Chapter 3 and 4 also, and the reader should refer back to these previous comments). Also, an abatement clause is included to remove all uses of a nonconforming character, i.e., those not conforming with the planning ideal. Empirical evidence developed earlier facilitates a critical review of zoning practices in Spokane and enables conclusions to be drawn concerning the ability of business to relocate in the long run under impact of highway improvement within the restrictions of planning policies and practices. The review is lengthy. This is because the topic is of great importance, and it is desired to present as complete a contrast as possible of present planning practices and concepts and empirical evidence developed in previous sections of the study.

Zoning for Business, 1929-1958

Zoning restrictions in Spokane were first imposed by the zoning ordinance of 1929. With few modifications this ordinance is still in use to this day. New business was intended to be directed to selected locations by prior classification of land according to acceptable use. No powers of removal of existing uses were granted. Zones were defined upon existing patterns of land use and could be altered where sufficiently strong cases for rezoning or variance were presented to the City Commissioners.

In general, retail and service businesses were permitted in three districts (zones):

1. Class III, Local business district. In this district were permitted all types of residential land use and retail and service establishments needed to serve demands of residential districts. However, bars were specifically excluded.
2. Class IV, Commercial district. All retail and service uses were permitted in

addition to all types of food processing, wholesaling and warehouses, storage and contracting facilities.
3. Class V, Manufacturing district. All types of uses permitted in the Class IV district, and all except noxious manufacturing uses were allowed in this zone.

Considerable latitude existed in the locational choice of firms. The City Plan Commission's "Land Use" report (1954) noted that out of the total area of the city of 26,170 acres, 2,698 were zoned Classes III, IV, and V. Of the acres zoned, only 1,392 were used for business, commercial, and industrial uses, and 433 acres were in business, commercial, and similar uses in the restricted industrial zone. Only 535 acres were used for business and commercial uses. More than 7,900 acres of the city were vacant. Also, until recently it has not been difficult to present an acceptable case for a variance or rezoning especially where several possible alternate sites were available.

These factors have meant that, as far as Spokane is concerned, existing zoning has allowed a spatial structure of business to develop under the operation of relatively free market forces.

Recent Attempts to Control Shopping Developments
In the past five years the city has pursued a more active policy towards business. This has consisted both of attempts to direct new shopping centers to prescribed locations by restrictive granting of new zones and variances, and also of development of a new proposed zoning ordinance to replace that of 1929. The former controls have been undertaken as integral parts of the latter planning and provide good indications of the direction of policy once it obtains more effective weapons in the form of the new ordinance, for premises used in the formulation of the proposed ordinance have been utilized in recent planning decisions and provide a clear picture of how the Plan Commission will operate.

How effective will these controls be? If implemented as planned, efficient relocation *in* existing zones, and compatible with the conceived structure of zones, will not be impeded. But existing nonconforming uses would be abated, and *new* construction, relocating under the impact of long-run changes in the structure of the economic system, will be made to conform strictly to the planning ideal. It is in this latter case of institutional restriction that the review here is of interest and value.

The proposed ordinance and recent policy both assume (after the work of the Urban Land Institute and the American Institute of Planners) that there should be two types of shopping centers, neighborhood and community, based upon two types of goods, convenience and shopping, with two types of associated trade area. Land has been designated for either neighborhood or community business center use and trade areas have been estimated on the assumption that neighborhood centers should be approximately one-half mile apart and community centers one and one-half miles.

With these assumptions, land required within each trade area for business has been estimated by the Plan Commission in the following manner. From the population map estimate the total number of families in the given trade area, and from census data estimate their aggregate income. Assuming that 50 per cent of incomes is spent upon goods and services (one-third in the central business district, one-third in community centers, and one-third in neighborhood centers), estimate total funds to be spent in the type of center being studied. Assuming that $50 per square foot of retail floor area is required for a business to be in the black, estimate the gross retail floor area the assumed trade area will support. To this add various allowances for offices, gas station, parking (at the ratio of 3:1 square feet for stores, 2:1 for offices), walks, service areas, landscaping, and so forth, and a total area of land to be used for business is found. Subtract the area of present uses (except those designated as nonconforming and therefore to be abated); there remains the allowable land for present business expansion. By using population forecasts, further amounts of land needed for expansion at successive future time periods may similarly be obtained. Rationales for the estimating ratios used are found in the works of the City Plan Commission.

If a proposed business center development satisfied a need for further business as estimated in the above manner and a location was proposed compatible with the proposed zones, then approval of the City Plan Commission could be obtained. If these requirements were not satisfied, the proposal was not supported (however, later approval could be obtained from the City Commissioners who have the veto power over the Plan Commission).

The Proposed Zoning Ordinance

Critical assumptions built into the proposed zoning ordinance are those which classify the business centers of the city into community and neighborhood centers and those leading to calculations of needed amounts of land based upon assumed trade areas. Associated with these assumptions and restrictions are decisions concerning size and spacing of centers and abatement of uses which are considered nonconforming.

Retail and service business establishments are to be allowed differentially in several zones within the city:

1. B1, Local business zone. This zone provides for small centers which contain retail and personal service establishments serving daily needs of a population residing within one-half mile of the business center. Only uses which serve the above purposes without detrimental effects upon surrounding residences are to be permitted. The location and quantity of land in a B1 zone is to be "commensurate with the purchasing power and needs of the present and potential population residing within the neighborhood trade area." Bars are specifically excluded. All uses have to satisfy certain specific site requirements. A list of uses allowed is as follows: apparel, art studio, bakery at retail, barber, bicycle sales and repair, confection, dressmaking and tailoring, drug store, dry goods, dry cleaner's pickup, fix-it, florist, fountain (not drive-in), frozen food locker, gift, grocery, fruit and vegetable, hardware and paint, hobby, interior decorating, jewelry, laundry pickup, meat market (except fish), music store and studio, newsstand, offices employing fewer than five persons, photo studio and camera shop, parking lot, public service building, radio and television repair, restaurant (except drive-in, dancing, and entertainment), gas station (site requirements restricted), shoe sales and repair, sporting goods and variety. Auto repairing is not permitted.

2. B2, Community business zone. This zone is intended to accommodate and con-

trol larger shopping developments with community facilities serving several square miles of the city, comprising a group of neighborhoods within a distance of one and one-half miles of the B2 zone. Goods provided and land zoned are to be commensurate with purchasing power and needs of present and potential populations residing in the assumed trade area. Location is to be convenient to the city's arterial system. Permitted uses include all those of the B1 zone plus amusement park, appliance, bank, bath sales, bowling alley, building supplies, business college, catering, commercial artist, department store, drive-in, eating establishment, employment agency, film exchange, floor covering, funeral home, furniture store, grocery and supermarket, hotel, leather goods, lithographing and addressograph, office and office supplies, massage, motorcycle sales and repair, nursery and greenhouse, open lot sales, optician, pet shop, public utility building, second-hand store, gas station and fuel oil sales, studio, theater (except drive-in), and motel (by special permit).

3. Other zones. Retail and service businesses of the above types will also be able to locate in the B3 central business district zone, the C1 commercial zone, and the M1, M1-L, and M2 industrial zones. In these various zones all retail and service business types excluded from the B1 and B2 zones are able to locate as well.

4. All other zones. The proposed ordinance also makes provision for abatement of all nonconforming businesses located in other zones. In all, 217 are identified, largely isolated establishments drawn from groups A and B (see Table 4-II).

The Planned Pattern of Business

The Plan Commission of the city of Spokane conceives a business structure for the city comprising neighborhood centers providing convenience goods and community centers providing convenience and shopping goods. Allowed uses reflect definitions of what constitute convenience and shopping goods. Amounts of land are determined as above. The whole system is to be oriented about the specialized central business district of the city. All uses in other than planned locations and of other than specified kinds in planned centers are to be abated. An actual trade area pattern is, by planning action, to be made to conform to the given pattern utilized in the analysis of land requirements and location of centers. Newly locating or relocating business must likewise conform to this pattern.

Critical elements of this planned pattern are stratification into neighborhood and community centers, specification of permitted convenience and shopping uses, and abatement of nonconforming businesses. Empirical evidence produced as a result of the Spokane study enables an objective evaluation of each of these critical elements.

Critique of Spokane Planning for Business

The nature of the empirical evidence developed in Chapters 3, 4, and 6 indicates not only that simple assumptions about types of goods and trade areas are in error but that out of the shopping needs of the city has emerged a far more complex structure of business than the Plan Commission imagines. There are three levels of a nucleated shopping hierarchy below the central business district rather than two (neighborhood and community). Also, there is an arterial hierarchy for which there is no planning provision at present, and an automobile row is not conceived. It may be entirely possible for the Plan Commission to continue with its present policy and ultimately force business into the pattern it desires, a pattern other than that which has developed as a part of the natural interactions of demand and supply in the city. However, to do so would be to risk incurring substantial social losses to the consumer, to the businessman, and to the city because of suboptimization of the system. Also, the full array of benefits from highway improvement will presumably be unable to accrue. It is the purpose of the ensuing remarks to indicate the limitations of present policy and to suggest means whereby this policy may be improved.

Permitted Uses

Planning policies assume two types of good, convenience and shopping, and broad groups of uses permitted in B1 and B2 centers are specified on this basis. When the empirical evidence presented in Table 4-II and 4-XIV is examined and contrasted with the broad use groups specified in Spokane, the inadequacy of the latter is clear. To be sure, they do not exclude anything of significance. It may be argued that it is the lack of differentiation between location requirements and habits which creates problems. Present planning concepts of what constitute convenience and shopping goods are quite inadequate.

The distinction between nucleated shopping and arterial functions is not recognized. Yet this basic dichotomy is not without importance for the planning of business since one type has one set of locational patterns, the other an alternate. Efficient planning should recognize and allow for such differential elements if business is to be kept healthy and the consumer to be served in an optimal manner and if benefits to arterial type businesses are to accrue from highway improvements.

Also, simple stratification of business into two levels, convenience and shopping, is far too diffuse. Groups of associated business types within both conformations have more limited and restrictive membership. It is possible, given the empirical evidence developed in Tables 4-II and 4-XIV, to prepare a much more precise plan for business at a greater series of better defined levels. This is possible not only for the gross pattern of business but also for both conformations.

An efficient approach to planning using natural spatial groupings of business to develop permitted uses would be to identify (1) two conformations of business types and (2) precise hierarchical stratification of businesses by spatial groups within each conformation. Recognition of these provides the first step. Also needed is information concerning location habits of the various resultant types of centers and their relations to purchasing power and trade areas.

Classes of Centers

The proposed zoning ordinance developed, using an assumption of two types of goods, a bimodal stratification of business centers into neighborhood and community classes. Just as the empirical evidence indicated the invalid character of the assumption, so it shows the lack of validity in the concept of two classes of center. If any bimodal character exists, it is between nucleated shopping and arterial types. The two have differing functional characters and differential location habits. Diffusion with orientation to home-based purchasing power and situation at intersections stratified according to traffic intensity are characteristics of the former. The latter is highly concentrated, oriented to major traffic arteries and to the demands of people moving. Also found here are supply and repair uses which provide their service in the home and are consequently located to be better able to reach the home. The former rely upon parking and pedestrian traffic between stores, the latter upon the single-purpose trip by automobile to the particular store upon the arterial. Parking is therefore a composite problem for the nucleated center and is a direct function of the individual store for the arterial type. The nature of hierarchical stratification likewise differs between the two conformations.

It is evident that not only should efficient planning take account of the differential natural spatial grouping of functions to prescribe permitted uses; it should also take account of both differential location habits and different patterns of hierarchical stratification. A modification of the present zoning system to recognize empirically valid spatial groups of business types would be adequate to account for the nucleated shopping facilities, but a *totally new concept* would be required to plan the arterial conformation, reserving stretches along, or set back parallel to, major traffic along frontage roads (see below) for these types of business and recognizing the essentially individual nature of the parking requirements of each type of store. Since it is evident that there is little functional relationship between the two conformations of business, both as regards shopping trips and location of purchasing power, an efficient policy could without social loss maintain a dichotomous planning and zoning structure based upon the conformations. Only with these recognized and permitted in the institutional framework can it be hoped that the full array of highways benefits will accrue.

Trade Areas

Calculations of land requirements based upon assumed trade areas are suspect. Purchasing power is not a single aggregate to be parceled between two types of shopping center and the central business district. There are at least two separate elements to the purchasing power aggregate. The first of these comprises expenditures made during shopping trips to one or other level of the nucleated shopping hierarchy, and these expenditures have to be shared between the various levels according to the amount of goods purchased in each. The second consists of a diverse group of expenditures upon goods and services associated with the automobile and its needs, with repairs, with furniture and supplies, and the like. This, too, must be parceled out according to the various purchases and the position of the serving business in the centers of the hierarchy.

This means that trade areas vary. The concept of a trade area defined upon a home base is adequate for the nucleated shopping type of business. However, it is doubtful whether it can be applied with equal validity to the arterial conformation. Here trade areas are likely to be multiple-referent composites of home-based, work-originating, highway-oriented, and business-service trips. In both cases prior assumptions about shape and extent of trade areas are of doubtful value. Certainly, if land requirements are to be calculated, the first need is a detailed trade area study both for the two conformations and also for *each* level of the stratified hierarchy within the conformations. From this kind of study desired information concerning total funds spent in each shopping center should be derivable.

Abatement

Suggested abatements would remove some 217 business establishments widely scattered throughout the city. These establishments are important parts of the lowest stratum of both nucleated shopping and arterial conformations; only 28 are business types of neither group A (gas, etc.) nor group B (grocery, etc.). It is by no means clear that removal of these scattered uses would lead to greater social gains, for they are integral parts of the complex system of complementary provision of goods and services to immediate residential areas at the lowest level. It is entirely feasible that many of them are necessary parts of the system, although tendencies toward centralization have been noted to result from highway improvement. Certainly, the question of abatement should be studied far more carefully in the light of the empirical evidence. Whether it is more desirable to force the lowest level constellation of complementary uses into larger centers or to allow it to remain locationally more diffuse certainly merits further detailed study. Without this further study any active policy of abatement, even though it may have immediate intangible merits, is questionable.

Some Measures of Gain and Loss to the City
Both tangible and intangible losses and benefits are likely to accrue from the direction of potential uses into suboptimal locations, or from abatement of uses which are constituent parts of an optimal system.

If the observed patterns of business are optimal, then restrictions which eliminate lower levels of the nucleated shopping hierarchy and which do not make special provision for the arterial hierarchy will result in tangible losses. The consumer will have to travel further to obtain his needs and therefore will have diminished purchasing power. Lower effective purchasing power means wider spacing of an optimal system of shopping centers. Uses of near-marginal, hierarchical character become marginal and submarginal and in the long run fail or have to seek new locations. Therefore, ability of businesses to compete for the land is reduced and general rent levels are diminished. The landowner suffers as well as the businessman. Also, institutional restrictions of this character inhibit the accrual of full benefits of highway improvement to the spatial system of business establishments and groups.

One of the major arguments for planning business locations is that many other tangible and intangible losses are avoided, or benefits thereby gained. For example, provision of special parking facilities and planned access greatly facilitates movement along arterials and reduces congestion. Planning of shopping centers is also desired to maintain community and neighborhood character, and by preventing cluttering and conflicts, such planning is held to prevent depreciation of residential land values in proximity to businesses. Included within the locational standards of the city of Spokane are several of these tangible and intangible factors not directly associated with the sale and purchase of goods and services but asserted to be responsible for either benefits or losses: compatibility with surrounding developments, assurance that the morals and well-being of residents will not be jeopardized, provision of safe access via a feeder street, assurance against overexpansion and loss of neighborhood character.

There is no explicit model of the residential sector of the urban economy which shows that uncontrolled developments can lead to losses in these several respects. But at the same time it should not be inferred that planning for business compatible with optimal location requirements is inimical to the maintenance of such standards if they are shown to be essential. This is best illustrated in the case of the arterial hierarchy, planning for which would be in such complete contrast to accepted concepts and practices. It would be entirely possible to plan for a string of arterial type businesses along a highway as a unit if the string of businesses were to be set back along a frontage road with adequate off-street parking and common controlled access from the arterial, so that congestion and the risk of accidents were reduced, yet with each business maintaining its apparently required visibility from the highway. Common site screening for the entire string could ensure against loss of neighborhood character and maintain land values of proximal residences. Recognition of the character and needs of the arterial hierarchy would be the assurance against overexpansion. Such planning would result in additional intangible benefits at the same time, for arterial business would thereby be kept out of nucleated centers and needless traffic through residential areas could be avoided by keeping arterial demands confined to the arterials.

There are both pros and cons, and discussion often resolves itself into assertions and counterassertions. Certainly, any planning decisions need to be based upon higher levels of *empirically* valid information than is at present evident in the work of the Plan Commission of the city of Spokane. But some limitations have been identified; and, at least intuitively, it seems that it is possible to plan for business centers located in an optimal manner and still avoid tangible and intangible welfare losses, and profit both from maintenance of location standards and beneficial safeguards and from the full accrual of the benefits resulting from highway development.

Long-Run Trends and the
Location of Arterial Type Business

Highway impact and consequent redevelopment of arterial-oriented business presents some perplexing problems made all the more perplexing because of the institutional problems above. It is well known that the location of arterial-oriented business on major thoroughfares produces congestion and safety problems. The strangling of major streets by business establishments has been a major factor in the need for new facilities, in the restriction of these new highway facilities to a limited-access character, and in the planners' rejection of strip developments of business. Yet these businesses are a convenience to the customer by being located on major streets. In particular, some of the goods and services supplied by arterial-located business are used in travel. The convenient gasoline station and lunch room are examples of business establishments of this type. Building materials establishments also tend to locate on arterials. The customer can easily find any of these when they are on arterial streets. Parking is simplified as are loading and unloading operations, and the establishment is in a convenient place to deliver goods and/or services to customers. Establishments of this type are not essential to the operation of the highway system, but efficient operation of the establishments requires that they be located on major streets.

The examples given above were of isolated establishments. Nucleated automobile row type associations are also found on city arterials, and the arterial location seems essential to their operation. Persons are willing to travel long distances to obtain automobiles, and location on an arterial running to or through a major business nucleation (hence consideration of group E as nucleated) supplies the easily accessible point with ample amounts of space for lots and parking. Here is another case where location on an arterial seems essential to the operation of a business, but the business is not essential to the operation of highway transportation over the particular street. Other aspects of automobile row locations were discussed in the previous section of this chapter. For example, distinctions have been drawn between highway-oriented and urban-arterial types of arterial business.

All types of arterial business locations are subject to relocation when the highway is improved. In the case where the establishment is essential to operating traffic over the facility, the need for relocation is obvious. In the case where location on arterials is essential to the operation of the business but the operation of the business is not essential to the operation of the arterial, the relocation force is more subtle. The establishments are at strategic places on the highway network in the first place. Reorientation of the highway network redefines what is meant by strategic places. The highway network is different from what it was before and relocation will occur in response to these differences. Because of the differences between the two types of relocations, they are discussed separately below.

Necessary Businesses

Businesses catering to through traffic have their well-being directly related to the amount of that traffic. In a gross way, the well-being of this type of business will certainly increase as a result of the accelerated highway investment program because there will be more traffic. Even if by decree no relocation of business of this type was allowed, their sales would increase because of greater gasoline consumption by the increased traffic and greater purchases of food away from home. This absence of relocation would mean disadvantages to the consumer, however, because he would have to travel out of his way to obtain these necessary provisions. In the presence of relocation, finding the highest level of benefits requires balancing conflicting goals. It is desired to convenience the consumer. This would require continuous gasoline stations and eating places along new facilities. Having least congestion would require having no gasoline stations and eating places, but neither of these extremes is feasible. Also, the number of establishments allowed affects the profitability of individual establishments, and this profitability must be considered along with the desire to avoid congestion but offer many services.

Very little evidence is currently available basic to the problem of the provision of these goods in the most satisfactory way (i.e., the way which balances the conflicting goals of continuous stretches or complete absence). The remainder of this section reviews fragmentary evidence on this point.

Locational Opportunities

Locational opportunities are limited by engineering characteristics of modern highway systems which in turn bear on the desire to avoid congestion as much as possible. Good information is available from Leisch (1958) as well as from many other authors on possible ways to provide access to the limited-access system. Opportunities are provided at frontage roads, interchanges and their lateral and transverse distributors and variants, and along parallel distributors. For the through traveler it is essential that business establishments be identifiable from the highway either by being in view or being well signed. Also, it is desired to avoid interrupting the through trip unduly. For these reasons, the frontage road would seem to be the most desirable location of business for the through traveler.

For trips of short duration or at the beginning or end of a longer through trip, these business needs and services of highway transportation may be met by locations at interchanges and on distributors. Problems of convenience for the traveler are again pertinent. Also, engineering these facilities for a free flow of traffic may conflict with allowing turn-offs and turn-ons related to business.

Demand for Goods and Services

Information as to what types of goods and services will be demanded by travelers on highways in the future is very slim. Experience with modern highway facilities has been largely limited to toll roads on which characteristics of trips undoubtedly differ from those on free roads. Also, businesses offered are by franchise. An example of information on demand by travelers is from a recent study by the Ohio Department of Highways (1957), and their findings regarding the use of facilities are summarized below:

1. On the average trip (70-90 miles) on the Ohio turnpike about half the motorists stop at plazas. On trips over 200 miles practically all cars stopped during the trip;
2. Of the cars stopping at plazas on the average trip 34 per cent buy food, 28 per cent use restrooms, and 21 per cent buy gasoline; only 3 per cent stopped for other purposes;
3. On trips up to 200 miles 75 per cent of motorists who stop buy food, and as the length of trip increases the proportion purchasing gasoline increases;
4. Few motorists leave the turnpike--on trips of up to 225 miles only 2.5 per cent of the motorists leave for lodging, 1.5 per cent for fuel, and only 0.8 per cent for food. Once on the highway, the motorist tends to remain on it until his trip is completed.

These facilities are of the type used as examples in the discussion just preceding and are of the highway-oriented group A types (Table 4-II).

Another source of information on the demand for goods and services is the market itself. One of the authors has a study in progress of evolving patterns of business location at interchanges and frontage roads on modern free expressways. The information is of such short-run standing that it is difficult to think of it as properly mirroring the demand for goods and services. The study area is a segment of U.S. 99 between Portland, Oregon, and Seattle, Washington. In particular, the area deals with a hundred-mile section of U.S. 99 between Tumwater and Kelso, Washington (Figure 6-1).

The study does not allow statements in the form: so many miles of highway with so much traffic require so many establishments of various types. This is in part because the pattern of location is not yet evolved and in part because the highway has establishments on it that were there prior to highway improvement (one intersection has a greenhouse, for example), and no operational way is available for

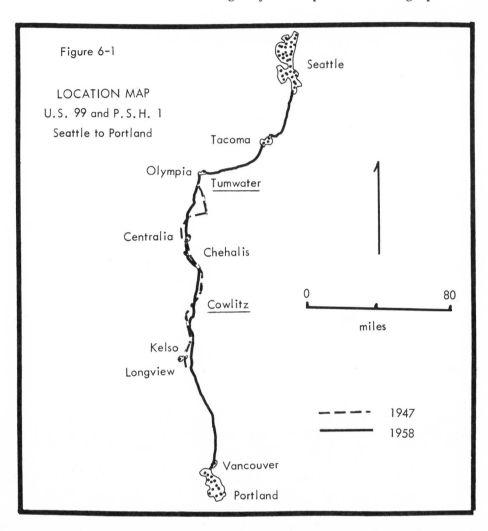

Figure 6-1

LOCATION MAP
U.S. 99 and P.S.H. 1
Seattle to Portland

defining whether or not a facility is on the highway (at an intersection, for example, how far out on the intersecting road does one go before deciding that an establishment is not associated with the through facility).

In a more refined way, it would be dangerous to assume that what was characteristic of this stretch of highway and this amount of traffic would be characteristic of another highway of the same length with the same amount of traffic. Traffic streams differ from place to place in characteristics of trips as regards the utilization of facilities, and no information is available on the particular characteristics of this highway utilization.

Fragmentary information is available on the evolving relocation pattern from the examination of actual developments. Two intersections are used here as example cases in point. One is the Tumwater intersection at the north edge of the study area (Figures 6-2, 6-3) and the other is the Cowlitz intersection toward the south

TUMWATER BEFORE HIGHWAY IMPROVEMENT

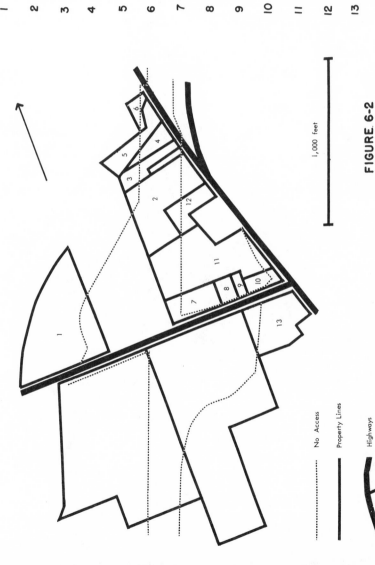

No Access
Property Lines
Highways

1,000 feet

FIGURE. 6-2

1 RESIDENCE
2 RESIDENCE
3 RESIDENCE
4 MOTEL
5 RESIDENCE
6 SHOP
7 JUNK SHED
8 RESIDENCE
9 RESIDENCE
10 GASOLINE STATION, STORE, GARAGE, BARBER SHOP
11 RESIDENCE
12 RESIDENCE
13 TAVERN AND CABINS

TUMWATER AFTER HIGHWAY IMPROVEMENT

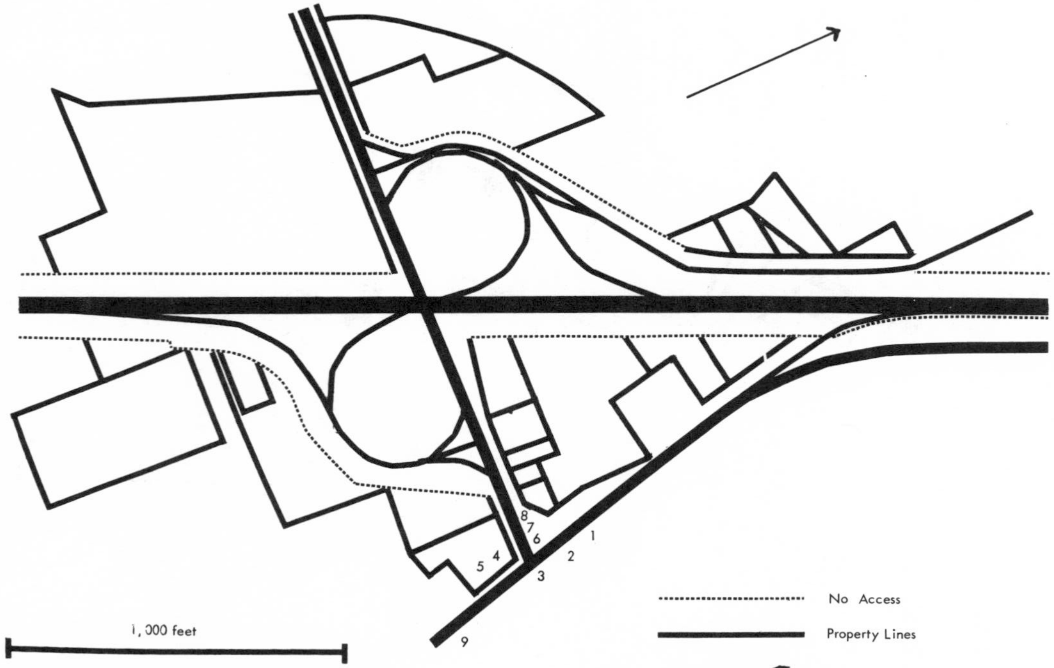

1 REAL ESTATE

2 BARBER SHOP

3 HAMBURGER STAND

4 TAVERN

5 CAFE

6 GAS STATION

7 GROCERY

8 BUILDING SUPPLIES

9 MOTEL

.... No Access

——— Property Lines

Highways

1,000 feet

FIGURE 6-3

Note: Partial information, highway is under construction

COWLITZ RIVER AFTER HIGHWAY IMPROVEMENT

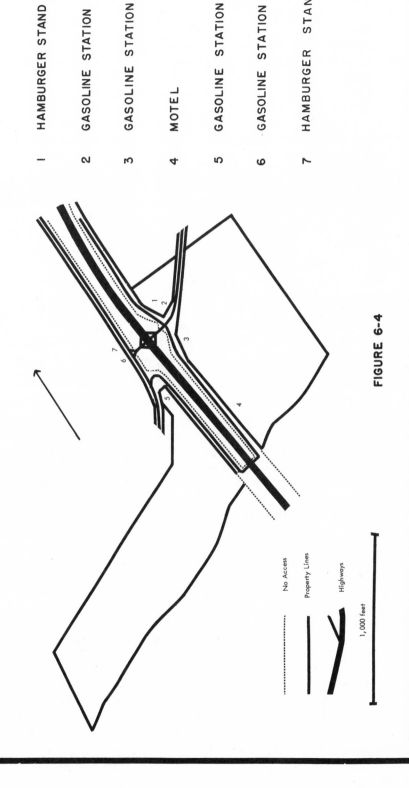

1 HAMBURGER STAND

2 GASOLINE STATION

3 GASOLINE STATION

4 MOTEL

5 GASOLINE STATION

6 GASOLINE STATION

7 HAMBURGER STAND

No Access

Property Lines

Highways

1,000 feet

FIGURE 6-4

edge of the area (Figure 6-4). The Cowlitz intersection also includes a stretch of frontage road.

Tumwater intersection lies at the northern end of the operating portion of the new U.S. 99 just south of the city limits of Tumwater, an expanding suburban residential town (population 1950, 2,725; population 1958, 3,667). All traffic moving north to Olympia and Tumwater leaves the new highway and joins the congested old highway at this point. Figure 6-2 describes the intersection and adjacent areas prior to first land acquisitions for the new facility in 1954. At the old intersection were located a gas station, garage, general store and barber shop, tavern and cabins. Nearby were a junk dealer, a motel, and another shop. There were also residences here.

Land acquisitions began in 1954 and continued through 1955. Construction of the new highway is as yet incomplete, although the figures note the entire right of way. Today southbound traffic moves along the new highway, and northbound traffic moves along the old highway from the intersection. The old highway will remain in use as part of the intersection after completion of the highway for northbound through traffic, but traffic flows will be reversed.

Construction of the new facility has meant the elimination of older land uses in the right-of-way, viz.: motel, shop, and junk yard. It should be noted that more will be removed (Figure 6-3), notably a gas station, grocery, and building supply facility at the old intersection, and now lying in the right-of-way. But new land uses have come to the old intersection: cafe, real estate, and hamburger stand. There will remain one barber and one motel, but the sites of these facilities are no longer the same. Land uses with respect to new accessible locations around the new intersection are unknown.

The North Cowlitz intersection is located in proximity to the declining lumber town of Vader (population 405). There, in 1956, an old inefficient intersection was replaced by a more modern facility to be of limited-access character when an overpass is constructed. The site provides substantial frontage-road opportunities for new business. Before the new facility no business land uses were found at the intersection or on the frontage road. Since 1956, however, several new businesses have been opened, viz.:

1956: 3 gas stations, 1 motel
1957: 1 cafe
1958: 1 gas station, 1 hamburger stand

These businesses are located where access is available but are set back from the intersection and along the frontage road.

It is evident that the short-run changes are these: elimination or relocation because of the taking of right-of-way or denial of existing access; definition of new site possibilities; growth of highway-oriented uses located with respect to the highway but defined on the new set of locational opportunities. Short-run growth is particularly noted in the always arterial highway-oriented category (Table 4-XIV).

The fragmentary character of the evidence has been emphasized throughout this discussion of the impact of highway improvements on business *directly oriented* to the highway traveler. It is difficult to see how the impact could be anything other than favorable for operators who are alert to new location opportunities afforded by highway improvements and who capitalize on them. The increased traffic concurrent with better highway facilities assures favorable developments. On the other hand, it has been emphasized that the benefit from highway impact must take into account convenience to travelers, congestion, and the operating characteristics of types of business. Very little information is available on these factors although significant work has been done and is continuing. The opportunity is available for favorable expansion of business. The degree to which opportunity will be realized depends on factors not yet completely evaluated and proper action on the basis of these factors.

Unnecessary Establishments

Many of the establishments now located on arterials are not essential to the operation of the arterial as a traffic facility although the existence of the facility is essential to their effective operation, and these include both highway-oriented and urban-arterial types noted in Table 4-XIV. It was noted that the location structure of business is subject to change because of highway change. The problem now is the nature of location change so far as these establishments are concerned. It is presumed that any improvement in transportation will affect these establishments favorably by increasing their market areas and increasing efficiencies of operation. What is unknown is the extent to which this will cause relocation and the exact character of this relocation.

Opportunities for relocation seem best at interchanges on the interstate system. These will be readily distinguishable as locatable places on the system and as areas where amounts of land are available and suitable to the requirements of these facilities. It is desirable that these facilities be at such locations because it conveniences their effective operation. Their location at interchanges may create congestion, however, and lead to undesirable effects. Other than these general observations the authors have no empirical bases for statements on these kinds of businesses.

Long-Run Trends and the
Location of Nucleated Type Business

Problems are perhaps less perplexing when questions concerning highway impact and relocation of nucleated type businesses are considered. For one thing, present institutional structures of planning and zoning are designed to accommodate nucleated centers, and zones are defined so loosely that few restrictions upon the character of centers at each level of the spatial structure are to be expected. Also, the tendency for development of new nucleations as planned shopping centers has become well established since the Second World War. More than 2,500 planned shopping centers have been constructed in the United States in the past fifteen years, and these, although varying in size, maintain the character noted in several sections of Chapter 4. Along with planning has come controlled access and egress, adequate off-street parking, and astute, even though intuitive, location with respect to expanding urban markets for goods and services provided in nucleated shopping centers. It will be noticed, however, that these planned nucleations exclude the automobile row type of facilities.

On the other hand, many perplexing questions still remain. Location of planned centers is not essential to the operation of the highway system (except to the extent that they generate traffic), but location of centers with respect to the system plays a vital role in the success of centers. Each center at each level is best located central to the maximum profit area at its command; centrality implies accessibility to residences via the highway system. For lower order centers accessibility may be only by local streets; for higher order facilities people will travel further and will seek to move with greater facility via the higher quality streets and highways available. Tendencies to centralization have already been noted in the short run (as in the case of Marysville). These tendencies have been associated with improvements in highway facilities which enable customers to move longer distances to more specialized facilities in higher order centers. But there is only limited evidence concerning the long-run implications of these tendencies in terms of the location of new shopping centers (as was noted previously, the tendency today is for new nucleations to be developed as integrated shopping centers), and the succeeding discussion of new locational opportunities provided by modern highway systems is consequently strictly intuitive.

Locational Opportunities

Problems associated with engineering characteristics of modern highways have been presented earlier. What is evident, however, is that centrality to surrounding populated areas, or to access via the modern highway and the surrounding areas, is provided only by interchanges. Frontage roads do not have the desired qualities of centrality possessed by interchanges--centrality which is essential to the operation of nucleated business. Leisch (1958) has reviewed a variety of possible shopping designs with controlled access at interchanges, and these need not be repeated here. Since evidence is so slender, little can be said about nucleated business types or classes of nucleated centers locating differentially at interchanges, although use of the Brink and de Cani model outlined in Chapter 3 with the interchanges as the set of possible center locations could presumably be of value in the study of actual situations where other factors so essential to understanding of the situation, such as purchasing power and population distribution, are known.

Centralization

Demands for locations and types of shopping centers will also change because of the increasing centralized specialization of the spatial system as people are able to move faster and more cheaply and easily. The greater the accessibility of a location, concomitantly the greater its desirability as a specialized central location for provision of nucleated shopping facilities.

Orientation and capacity of links on the highway system are extremely important variables in the process of the centralization of business. Characteristics of the system will determine whether change will take place by rearranging business establishments within present centers or, at the other extreme, whether a system with centers at new locations will evolve. Without doubt, there will be business rearrangements of both types. A complicating feature here is the decentralization of market areas as residential areas decentralize with more transportation available. These notions imply that although business centers will be characterized by more centralized groups of business, centralization will not necessarily take place within present locations and the centralized pattern may be more widely dispersed than the present pattern. Change will vary depending on the nature and orientation of the facilities provided. One illustration of centralization in the system is presented graphically in Figure 6-5.

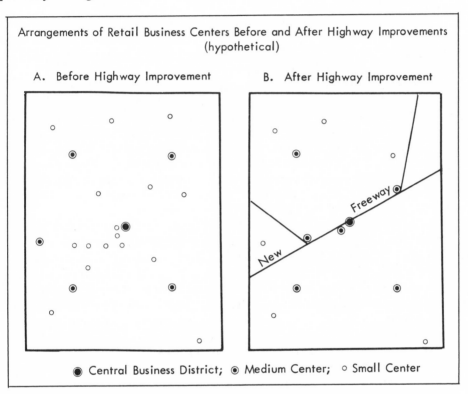

Arrangements of Retail Business Centers Before and After Highway Improvements (hypothetical)

A. Before Highway Improvement   B. After Highway Improvement

● Central Business District; ◉ Medium Center; ○ Small Center

Figure 6-5

Conclusion

The previous four chapters have inquired into location patterns of retail business and the impact of highway improvements on these location patterns. Almost without exception, location patterns are determined by the present network of highway facilities and represent the efficient arrangements of specialization owing to the availability of this transportation. The improvement of the transportation system will introduce further efficiencies into the location system for retail business; this is the impact of continued transportation development.

The nucleated system of shopping centers will be subjected to even greater centralization, smaller shopping centers will lose functions which can be provided more efficiently by larger centers but, in turn, will gain functions from still smaller shopping centers.

It is moot as to the exact location system that will evolve. This depends on factors such as present planning policies and practices, on the decentralization of the market with increased suburbanization which is of considerable importance for the location structure of business, on innovations in merchandising, and on the arrangement of the transportation system itself. A transportation system with the facilities oriented in terms of present land uses will tend to maintain the present location structure of shopping centers. A transportation system oriented in terms of presumably desirable reorientations of population and centers will lead to some other location structure. Transportation plans now available have elements of preserving present patterns as well as allowing some change. To determine exactly how these patterns will develop requires a greater range of information and a more refined theoretical point of view than that currently available.

Certain kinds of retail business are regarded as an essential part of the transportation function (e.g., the gasoline station). Surely the enlargement and reorientation of traffic streams will have a tremendous effect on these establishments, and the effect can hardly be other than favorable. The exact nature of the impact will depend on location opportunities afforded in the development of the highway system and the exact demand that materializes for goods and services. Some fragmentary information has been presented on patterns of relocation, and though there is no question of what impacts are present, the magnitude of impact can be estimated in only a fragmentary way because of a lack of basic information.

Some business establishments are oriented to major traffic transportation facilities, even though they are not essential to the operation of transportation on the facilities (e.g., sand and gravel sales). It is certain that improvements in transportation will decrease cost of consumer movement to these establishments and of the delivery of goods and services to customers of these establishments. In turn, these results mean a favorable impact. Here is another case where practically no information is available as a basis for an estimate of the ultimate effect of the impact.

Empirical information and a theoretical point of view have been provided in support of the above conclusions. A theoretical point of view stems from a body of concepts on the arrangement, spacing, and size and function of urban centers and shopping facilities; the empirical information was from several case study areas. Empirical information was provided not only from much previous work but also from studies of Spokane, Cedar Rapids, Cincinnati, Phoenix, U.S. 99 between Everett and Seattle, Washington, and planned shopping centers in various places in the nation. Information on change in structure was provided by study in the vicinity of Marysville, Washington, and in the cases of Tumwater and Cowlitz, Washington. While this is a massive amount of empirical information, the problem which it is desired to solve is very large and intricate. The information and concepts provided assist in understanding the character of highway impact. They do not provide a complete encyclopedia of information, but no one study ever will.

Succeeding sections and chapters in this volume continue themes identified in this chapter and provide new directions for research. In particular a great deal of attention is given to codifying relations between travel and specific land uses (Sections III, IV) and to rearrangement of the location structure in a particular case (Section V). It is easy to see how information of this sort is just as essential in dealing with the problem of impact of highway improvement now before us as are the materials presented here in Section II.

SECTION III

Highways and Urban Residential Land Use

This section deals with general aspects of the relationship of the urban road network to residential land use within the city. Chapter 7, "Spatial Aspects of Residential Land Use," presents the general problem of the spatial structure of residential land use in urban areas. The relationship of the spatial structure to the urban road network is pointed out through the examination of several general models of urban growth and structure, as well as examination of factors which enter into the processes of residential site selection by individual households.

The importance of access factors in determining the value of a residential site is pointed out in more detail in Chapter 8, "The Value of Access to Urban Residential Land," and the complexity of the measurement of access factors is discussed. Three empirical studies, one in Spokane, Washington, and two in Cedar Rapids, Iowa, which attempt to measure this relationship are also presented. These studies serve to point out the necessity of utilizing more complex models in attempting to explain the contribution of access to residential land values. The role of transport inputs to the household in determining the value of the residential site occupied by the household is also briefly discussed.

Chapter 9, "The Demand for Transportation at Urban Residential Sites," discusses previous theoretical and empirical studies on this topic and, using detailed information for a sample of households in Cedar Rapids, Iowa, presents some preliminary studies of the relationship between transport inputs to the household, location of the residence, availability of transportation, and the socioeconomic structure of the household.

7

Spatial Aspects of Residential Land Use

THE LOCATION of residential areas within a city is the result of a complex series of events which may be understood only in their relationship to the general urban rent system. The concept of a general rent system has been introduced and explained earlier in this volume (Chapter 3). Residential land use is a part of the general group of land uses bidding for favorable locations within the city, but it is not, under normal conditions, capable of outbidding most commercial uses and many industrial uses. However, through restrictions, especially zoning regulations, which have been placed in operation by our society, the bidding power of residential land uses has been artificially bolstered by reducing the competition for certain sets of locations.

The areas of residential land in a city grow and sustain themselves through a continuing process of site selection by individual households. The ability to answer questions pertaining to the impact of transport innovations upon the spatial structure of residential land use depends upon knowledge of the way in which factors such as location and accessibility enter into the site selection process.

The process of residential site selection by households is a complex one since there are many factors, both objective and subjective, which the householder must evaluate before his final choice is made. The accessibility of the site in relation to commercial, industrial, and other land uses; the character of the surrounding neighborhood; characteristics of the individual dwelling unit itself; and the cost of occupying the dwelling unit, are all factors of major importance in selection of the most favorable site.

The Place of Residential Land in the General Rent System

Although residences exert a great influence on the location of individual retail units and low order shopping centers, they are generally regarded as a city filling function whose location depends primarily upon the basic industrial-commercial spatial pattern. While this may be regarded as true in the short run, the total urban complex in the long run is an integrated system in which every part affects all others, so that the spatial structure of any given land use can be understood adequately only in relation to the spatial structure of the entire urban area.

Types of Residential Land

In the past many research workers have treated residential land as a homogeneous entity. Unfortunately, examination of the real world shows that this is not the case. The intensity of use of residential land has a direct bearing upon its ability to bid for locations. For instance, in competing for a given site it is doubtful that any single-family dwelling unit would be able to outbid a large apartment house. In this example, a difference in the purpose of the land use as well as a difference in intensity may be seen. The single-family dwelling unit acts as a home for one household. The householder in bidding for the site is faced with limitations imposed by the family income and expenditure structure, and is concerned with both tangible and intangible returns which the family expects to receive from occupying the site. On the other hand, the operator of the apartment dwelling is engaged in what is nothing less than a commercial operation, wherein he is concerned with cost of the operation and the returns on his investment. Because of the greater intensity of this land use, he is easily able to outbid the normal householder in competition for locations within the city.

Thus a major differentiation in the types of residential land may be seen, a differentiation based largely on the purpose of operation of the residential units involved and keyed largely to the intensity of site utilization. The discussion in this section is concerned with just one of the several types of residential land use--the single-family dwelling unit. It is hoped that further investigations will be made into the spatial structure of other types of residential land use.

Returns to Persons Utilizing Residential Sites

The returns to a commercial operator who is operating a high intensity residential unit as a business venture are quite similar to those obtained by operators of retail business. Returns to households utilizing residential sites are considerably more complex in nature. While the investment value of the residential site is of some importance to the householder in choosing a location, it is often not the major factor. The indirect returns he may expect to receive (i.e., reduced costs of movement and certain amenity and cultural values) are frequently items of some weight.

The type of returns normally anticipated by householders from a residential site are those of convenience of location in terms of the normal places of shopping, work, school, recreation, and so forth; returns from the character of the neighborhood (i.e., that the residents in the neighborhood will be persons of comparable socioeconomic status); and amenity returns arising from the nature of the dwelling unit itself and the surrounding land.

Bidding for residential sites is competitive and is conditioned upon monetary values. The amount that a given household is willing and able to bid for a residential site is conditioned in part by the total amount of resources which the household has available. Of equal importance is the current expenditure structure of the household. How much of its disposable income is the household willing to invest in housing? A recent study by Grebler, Blank, and Winnick (1956) seems to indicate that the proportion of the family resources that are available for the acquisition of residential sites has been declining. They state:

> There is at least a strong presumption that housing has suffered a decline
> in the consumer's scale of preferences, resulting from the newer goods and

services which have more successfully competed for a place in family budgets. The automobile, the growing emphasis on vacations and recreations, the popularity of "eating out, " movies, radio and television, and washing machines and freezers have profoundly affected the ways consumers spend their incomes. In this respect, housing--both an old good and a necessity--has shared the fate of other old and indispensable commodities.

Whether this decline in the preference for housing expenditures varies among households of varying socioeconomic status is as yet unknown.

Some Artificial Restraints and Their Effect on the Rent System
In a free and unrestricted bidding situation the relatively limited resources of most households would prohibit their competition with most commercial and many industrial uses for sites within the city. However, artificial restraints have been placed by society upon the system of bidding for locations, restraints whose purpose is to improve the competitive position of the householder. Most cities of any size in the United States today have zoning ordinances wherein certain areas of the city are set aside for use as residential sites, and other areas of the city are specified as locations for the wide variety of commercial and industrial land uses. Residential uses are not prohibited from occupying areas assigned to commercial and industrial uses, provided that they can effectively compete for the sites. However, the converse does not hold true since commercial and industrial uses are prohibited from bidding on sites located within areas designated as residential by the zoning ordinance. Thus, zoning ordinances tend to operate in favor of the householder who wishes to bid for a residential site. Under normal circumstances his competition is limited to other householders, presumably with the same or similar resource levels, and commercial and industrial uses are excluded.

Restraints of a private rather than a public nature are sometimes placed upon residential areas. These take the form of restrictive covenants which prohibit certain households from bidding on sites within the area. In some cases the covenants restrict the areas to households of a specific racial group or, more commonly, to households that are capable of constructing and maintaining a dwelling unit valued above some specified level.

Of the two types of restrictions on the system, the zoning restrictions are by far the more common and widespread. However, in practically any city there may be some small area or subdivision development wherein restrictive covenants will also be in force, thus further restricting the free operation of the general rent system.

The General Spatial Structure of Residential Land Use

Looking at the spatial structure of any urban area, we see major and minor commercial foci, industrial areas, areas of public land, and widespread areas of residential land. During the last half-century several theories have been set forth to account for the observed spatial structure of land use in American cities. Perhaps the earliest explanation was that set forth by Hurd (1911). He contended that the observed pattern of land uses in the city was the result of the growth process of the city, with growth taking place away from the point of origin in all directions along the lines of least resistance or greatest attraction or their resultants. According to this theory, central growth takes place from the heart of the city, and also from each subcenter of attraction, while axial growth pushes into the outlying territory along transport lines. In this structure of the city, residences are driven to the circumference of the growth area while business remains at the center. As growth continues, residences divide into various social grades, and low order retail functions move toward them with wholesale shops in turn following the retailers. Complicating this broad outward movement of zones, traffic arteries project shops into residential areas, creating minor business subcenters and thus tending to change the general circular pattern into a star-shaped pattern.

The Concentric Zone Theory

Burgess (1923) advanced the theory that, in the absence of counteracting factors, the modern American city would take the form of five concentric zones:

Zone 1: The central business district.
Zone 2: The zone in transition. An area surrounding the central business district which would include areas of residential deterioration caused by encroaching business and industry from Zone 1.
Zone 3: The zone of independent working men's homes. An area largely constituted of neighborhoods of second generation immigrant settlement in northern cities.
Zone 4: The zone of better residences. Residential areas of middle-class native-born Americans, small business men, and so forth.
Zone 5: The commuters' zone.

These zones were held to be a product of radial expansion taking place from the central area of the city. Burgess considered this radial expansion to be the factor bearing most heavily on city structure, although he admitted that the presence of topographic features and transportation arteries would introduce distortions into the zonal pattern. This model represents a somewhat more explicit and simplified statement of the ideas of Hurd.

The Sector Theory

Hoyt (1939), in a study conducted for the Federal Housing Administration, noted that within urban areas in the United States most of the residential areas tended to be distributed in a definite way with respect to the commercial and industrial districts of the city. He contended that the general spatial structure of cities had the character of sectors rather than concentric zones, with the sectors taking the form of wedges radiating out from the central business district along transport routes. According to Hoyt, different residential areas "tend to grow outward along rather distinct radii, and new growth on the arc of a given sector tends to take on the character of the initial growth in that sector."

Hoyt felt that the movement of the high rent residential area was the most important organizing factor in urban growth in that its movement tended to pull the growth of the entire city in the same direction. This high rent area, it was felt, tended to move forward from the center of the city along a specific avenue or radial line. New construction for the higher rent groups would be located on the outward edges of the old high rent area, and as the high rent areas grew outward, lower rental groups would move to occupy the housing which was formerly occupied by the higher income groups.

It is felt that the sector hypothesis tends to underestimate the importance of institutional controls such as zoning regulations in shaping the urban patterns, and that the emphasis upon residential land use as one of the key determinants in urban growth is as yet unverified.

The Multiple-Nuclei Theory

Harris and Ullman (1945) have offered a modification of the concentric zone and sector theories which they term the multiple-nuclei theory. This theory contends that in many cities the land use pattern is focused not solely around a single center (e.g., the central business district) but around several discrete centers. The rise of the separate foci for urban development reflects a combination of four factors: (1) certain activities require specialized facilities, (2) certain like activities group together because they profit from cohesion, (3) certain unlike activities are detrimental to each other, and (4) certain activities are unable to afford the high rents of the most desirable sites. The number of nuclei to be found in any given city is a result of the historical development of that urban area and the operation of localization forces within the city.

The general theories reviewed here are not a means to the actual and precise description of the spatial structure of land uses in any city but are more or less normative descriptions of over-all urban growth and structure. These general models are of some assistance to us in our search for information on the way in which transport innovations affect the spatial structure of residential land use, but the answer to more specific questions must be found in an examination of the way in which individual households choose their residential locations.

Factors in Residential Site Selection

In evaluating a given site as a possible residence location, there are two major groups of factors which a household usually takes into consideration. The first set deals with the spatial relationships of the site to the rest of the world, and the second set deals with the physical properties of the site itself. In deciding whether or not to bid upon a given site, a householder examines these factors and then evaluates the returns that his household expects to obtain from occupancy of the site. A monetary value is then assigned to the site, taking into account the family's income and expenditure structure. Having thus evaluated each of a given set of sites, the householder will presumably choose that one which will maximize the returns to his family. He may, of course, choose none of them, thus indicating that all assigned values fall below some minimum threshold level. An examination of the factors that may enter into establishing the level of returns to be expected from a residential site is now in order.

Access Factors

Since our society is a highly interdependent one, no family in an urban area today may obtain at its residential site all the things necessary for its existence. The members of the household must travel, and travel frequently, to other locations within the city in order to obtain both the sustaining and the social necessities of life. In order to minimize the time spent in this movement, the household should choose a residential site which has maximum accessibility to all activities in which the family engages. What are these activities? The following may be readily noted: shopping, work, school, church, and recreation, all of which are directly involved in the day-to-day operation of the family unit.

The complexity of the choice involved may be seen in the examination of just one set of locational reference points--those used in family shopping. The complex spatial structure of retail business in the city was pointed out earlier in this volume (Section II), and the householder must relate the location of the residential site under consideration to one or more members of each level of the retail hierarchy.

In addition to its location in relation to the activities in which the household takes part, the location of the residential site in relation to other land uses and socioeconomic groupings of persons is important. While a high degree of accessibility to many functions of the city is an asset to a residential site, close spatial proximity to certain functions may well tend to lower its value to the household. This topic is discussed at some length by Solow, Twichell, and Tiboni (1950) in their work on the appraisal of the neighborhood environment.

Urban sociologists, such as Hawley (1950), have noted in American cities a tendency toward segregation of similar family units. Those families with comparable socioeconomic characteristics seem to seek similar locations and consequently cluster together in selected areas within the city. This attraction of similar family units for one another in residential site selection has its basis, to a large extent, in a postulated uniformity of their location requirements. Families in the highest income brackets are able, through their greater ability to bid for sites, to have a fairly free choice among the available residential sites in the city. On the other hand, families in the lower income brackets face a more restricted range of choices in their selections of residential sites. Hawley also notes that the presence in an area of a given type of family unit is a localizing factor for other units of that same general type. The reason for this seems to be that a number of similar units can create by their congregation various amenities that are not inherent in location itself. If together in sufficient number, they can attract special services to their area, can engage in their own peculiar forms of selective behavior, and can, when necessary, offer relatively effective opposition to undesirable encroachment from without.

Site Characteristics

Under the general heading of site characteristics are included all factors pertaining directly to the site itself and not to its locational aspects. Among these might be listed such things as topographic configuration, size of the lot, and quality of the structure elected thereon. In reality the latter factor, quality of the structure, is not entirely independent of the locational characteristics of the site. However, for the purposes of this discussion it will be considered as a site characteristic.

The Impact of Highway Development

What is the impact of a change in the urban road network upon the spatial pattern of residential land use? As was pointed out in the discussion of the general models or urban land use, transport routes are the arteries along which urban growth takes place. The introduction of a new artery, of course, provides a new avenue for growth. However, transport improvements are often of a more subtle nature than can be seen in the introduction of a completely new route. Quite often they are concerned merely with changes in capacity and ease of movement along existing routes. The general models discussed earlier do not explicitly include any differentiation of transport routes by capacity or ease of movement. However, by a simple extension, we may include in these general growth models transport routes that are differentiated by these factors and, hence, will show varying effects upon urban growth. Thus, a subtle change such as capacity changes in a route may be seen to have a widespread impact on the general growth pattern of the city. However, because of their simplicity, these general models are unable to provide answers to specific questions regarding the impact of transport improvements upon the spatial structure of residential land use.

From the standpoint of the individual householder who is evaluating a site as a possible residence location, the urban road network serves as a connecting system between that site and all other locations within the city. The effects of a gross change in the urban road network, such as the introduction of an entirely new route, upon the access linkage of a given site to the rest of the city can easily be seen. But what about the more subtle change which may occur in the capacity and ease of movement along a route? There is reason to suspect that the desire to move along a route is directly related to the capacity and traffic volume on that route (see Chapter 9). Thus, a change in the capacity or ease of movement along a route between a residential site and some other location in the city may well change the demand for travel over that particular route. But again a more complex case arises. What of capacity changes in the urban road network which occur at places far removed from the residential site under consideration? The situation would be considerably simplified if one could assume that distant changes in the network would have no effect upon the accessibility of the site. However, Garrison and Marble (1958) and other authors have indicated that it must be recognized that capacity changes in any portion of a highway network may bring about changes in traffic volume in other sections of the road net. Thus, the more complex situation seems to prevail, and consequently, statements concerning the impact of a specific transport innovation upon the spatial structure of residential land use are difficult to make.

8

The Value of Access to Urban Residential Land

ASIDE FROM POSSIBLE modification of site characteristics, the urban highway system has its greatest importance to residential sites in the connection of those sites with other areas within the city. The importance of access factors in determining the value of residential sites has been noted by many students of urban structure. However, empirical examinations of the actual relationship between location and the value of residential sites have been infrequent. While most of the empirical studies which have been done concerned sites located in suburban or even rural areas, the similarity of the problems involved is such that their results provide useful guides for the present study of residential land values in urban areas.

It is the purpose of this chapter to attempt some simple empirical studies which examine the degree of association existing between the value of parcels of urban residential land and their location.

Previous Empirical Studies

The recent study by Garrison (1956) reviews many of the significant studies of the value of access to sites located in rural areas. This study empirically examines the value of access by various road types to parcels of farm and nonfarm land in three Washington counties, and it also represents the first attempt to define the location of a parcel of property in terms of multiple reference points in space. The study revealed variations in the contributions of the different road types to the value of the property; also, better results were obtained when distances were weighted by yearly trip frequencies than when simple unweighted measures were used. The theoretical implications of these findings are pointed out in another article (Garrison, 1957).

A study of residential satisfaction in the rural-urban fringe near Eugene and Springfield, Oregon, by Martin (1953) also examines the value of accessibility to residential sites. The author in this study is concerned with the extent of satisfaction among families living in the rural-urban fringe. A "rural-urban preference scale" was utilized to measure satisfaction with residential location. As a result of this study, it was concluded that while the extent of accessibility of the residence to the city center might well be important in certain individual cases, in general it was not a crucial factor in determining satisfaction with residential location in the fringe area.

In summarizing the study, Martin emphasizes his agreement with other statements of the importance of accessibility in the day-to-day operations of the family, and agrees that it is as important a factor to fringe residents as it is to families in other parts of the urban community. However, while emphasizing its importance, he feels that the degree of accessibility of any given residential site need not be high, and that other factors can and do compensate for the irritations involved in occupying a relatively inaccessible location. General location theory would indicate that one of these compensating factors is a decreased value of the site brought about by the substitution of transport inputs for rent payments (see Isard, 1956). Variations in the way in which different households recognize distance factors may also be seen in Martin's statement that "distance is always relative, and is always experienced in terms of the cultural patterns and previous conditions of the person involved."

Accessibility and Urban Residential Sites

Before embarking upon any empirical study of the value of location to residential sites, it is necessary to specify how the location of the site is to be defined. Location is relative and may be measured in relation to any number of other locations. Therefore, from the standpoint of simplicity of operations, it must be decided what other locations are of the most importance to a household occupying a residential site, since it is in terms of these locations that the evaluation of the location of the residential site will be made.

Accessibility to Commercial and Industrial Foci

Within the city many commercial and industrial foci exist, and it has been shown empirically that a hierarchial structuring of the commercial foci exists (see Chapter 4). The most frequently occurring commercial centers are those in the lowest level of the hierarchy, and each higher level is represented by centers which, while fewer in number, are of greater functional complexity.

Each level of this hierarchy is characterized by a specific functional structure which is related to the varieties and types of goods available at that level of the hierarchy. Therefore, it may be postulated that commercial centers of the same level show no essential differences in the goods and services provided. Thus, the accessibility of the residential site may be examined in terms of its relationship to some member of each level of the hierarchy rather than to all commercial foci within the urban area. The commercial and industrial areas of the city are also the major places of work for members of the populace, and the orientation of the residential site to the place of work must also be considered. Therefore, for the first set of factors which might be considered in a simple model, the location of a residential site may be defined in relation to the various levels of the commercial hierarchy, as well as to the place of work of persons residing at the site.

Accessibility to Noncommercial Foci

Shopping and work are not the only off-site activities that occupy the household. Persons spend significant amounts of time visiting noncommercial sites in order to obtain things which are in essence noncommercial. Some of these noncommercial sites include schools, churches, homes of friends, and recreation areas. Under recreation areas, such things as public open places and playgrounds are included, rather than functions such as theaters, bowling alleys, etc., which are in reality commercial activities. Thus a second set of reference points, noncommercial in nature, may be included in the definition of the location of a residential site.

Owing to a lack of time and resources, the following empirical studies are limited to examining the value of a site solely in terms of its accessibility to the commercial foci of the city. It is hoped that further research will be conducted to examine the contribution of the noncommercial foci.

A Simple Model of Residential Land Values

Perhaps the simplest model that may be postulated of the relationship of access to residential land values is one in which every unit of distance is considered to be the same as every other unit. Using this assumption and the multiple reference point concept developed earlier, a model of the following form may be postulated:

$$Y = \beta_1 x_1 + \beta_2 x_2 + \ldots + \beta_n x_n$$

This is a linear model wherein the *betas* are the weights attached to each unit of distance, the x_n are the distances to the n reference points of the residential site, and Y is the value of the parcel of residential land. Thus for a sample of residential sites a value of Y and a value for each of the x_n may be empirically observed for each of the sites. The question then arises: how much of the observed variation in the values of the residential sites can be explained in terms of their different locations as measured by the various values of the x_n for each site? It is also desirable to know just how much each unit of distance, in terms of each of the x_n reference points, contributes to the value of the site.

The answers to these questions may be readily obtained through standard techniques of multiple regression analysis. This technique of statistical inference permits answers to be found to those questions which have been raised and specific levels of confidence to be assigned to these answers. The adaptability of this measurement model to high speed digital computers also facilitates the solution.

Residential Land Values in Spokane, Washington

Earlier it was noted that the shopping centers in the city of Spokane formed a structure having eight types, including the central business district (see Chapter 4). As a preliminary attempt at analysis, it was desired to see what relationship, if any, existed between variations in levels of residential land values and the location of the respective residences in relation to these eight types of commercial nuclei.

The information from the 1950 Census of Housing provided the most readily available information on residential land values in Spokane. The area lying within the corporate limits of Spokane is broken up into over 3,500 smaller subareas or "blocks" by the United States Bureau of Census for the purpose of reporting selected housing data. A random sample of 50 of these blocks was selected for use in a pilot study, and a complex measure of the location of each of the blocks was made in terms of its relation to the different levels of the commercial hierarchy of the city.

Simple airline distances were measured from each block to the nearest center of each level. Also it was noted if the block lay to the north or to the south of the central business district. In addition to the locational measures, two environmental measures were also tabulated: (1) if the census reported that the block contained dwelling units having more than one and a half persons per room (a measure of overcrowding), and (2) if the block reported dwelling units containing nonwhite residents.

Standard multiple regression techniques were utilized to examine the associations of the parameters in a simple linear model. The following estimating equation was developed:

$$Y = 865.9 - 4.68X_1 - 24.39X_2 - 36.37X_3 - 72.22X_4 - 12.00X_5 - 0.71X_6 -$$

$$12.98X_7 - 38.26X_8 - 75.86X_9 - 507.79X_{10} - 93.52X_{11}$$

where:

X_1 = the airline distance in miles from the central business district
X_2 = the airline distance in miles from the nearest III_A center
X_3 = the airline distance in miles from the nearest II_A center
X_4 = the airline distance in miles from the nearest I_A center
X_5 = the airline distance in miles from the nearest IV_B center
X_6 = the airline distance in miles from the nearest III_B center
X_7 = the airline distance in miles from the nearest II_B center
X_8 = the airline distance in miles from the nearest I_B center
X_9 = 1 if the block contained dwelling units reporting more than 1.5 persons per room, and 0 otherwise
X_{10} = 1 if dwelling units in the block reported nonwhite residents, and 0 otherwise
X_{11} = 1 if the block lay north of the central business district, and 0 otherwise
Y = average value of single family dwelling units in the block, in tens of dollars

The model may be considered only moderately successful. Examination of the results disclosed that the model had a coefficient of determination (R^2) of 0.40. Thus, 60 per cent of the observed block-to-block variation in residential land values remained unexplained, and of the eleven variables considered only one, X_{10}, was significant at the 5 per cent level. One other, X_6, was significant at the 10 per cent level, while two others, X_2 and X_4, were significant at the 20 per cent level.

From the above it can be seen that the relationship between the location of a residential site and its value is not a simple one. In this respect the results of this pilot study confirm the results of an earlier study of rural property values (Garrison, 1956) wherein simple linear distance measurements also failed to provide a satisfactory explanation of observed variations in land values. In this study of rural land values, it was found that a more complex measure of distance, i.e., the actual linear distance times the average trip frequency, proved to be a more successful measure. However, the operation of this particular weighting process requires the collection of a great deal of detailed travel information. If some simpler weighting scheme could be made operational, the complexity of the situation would be considerably reduced.

A More Complex Model of Residential Land Values

The model examined in the previous section failed to provide as clear and concise a statement of the value of access to residential property as might be wished. There are at least two ways in which the simple model may be made more realistic. First, the use of airline distances is somewhat unrealistic since people do not move from place to place in this fashion. It seems probable that distances as measured over the road network of the city would provide a more satisfactory criterion. Secondly, aside from the trip frequency weighting discussed earlier, it has long been held that some system of weights must be attached to the distances so that the first unit of distance will not be counted in the same fashion as the last. Some form of inverse distance relationship has long been popular with students of spatial interaction. The work on movement of persons between cities by Iklé (1954) provides an example of some of the more common weighting techniques. Considerable controversy has existed over the power coefficients to be attached to the variables in this type of model. It is possible to obtain one of the simpler forms of weighting by performing a logarithmic transformation upon the previous model to obtain the form

$$\log Y = \log \beta_1 x_1 + \log \beta_2 x_2 + \log \beta_n x_n$$

which in its untranslated form is:

$$Y = AX_1^{\beta_1} X_2^{\beta_2} \ldots X_n^{\beta_n}$$

This latter nonlinear form is similar to the economist's Cobb-Douglas production function. It is proposed to fit a model of the above form in the hopes of obtaining a better explanation of the variability of the values of residential sites.

Residential Land Values in Cedar Rapids, Iowa

The city of Cedar Rapids, Iowa, was chosen as the site of the second study of residential land values since (1) it was the only other city in which an empirical study of the structure of the retail hierarchy had been made, and (2) a considerable volume of travel data on individual households was available if needed (see Chapter 9).

Cedar Rapids is a city of some 80,000 persons located in eastern Iowa. The generalized land use within the city is shown in Figure 8-1, and more detailed comments upon the commercial structure of the city may be found in Section IV of this volume. The general distribution of households within the city may be inferred from the generalized distribution of dwelling units shown in Figure 8-2.

The travel sample in Cedar Rapids consisted of 100 randomly selected households. For the purpose of comparison, the land value sample consisted of the blocks in which these 100 families resided. After elimination of unreported blocks and families with poor travel information, a sample of 71 blocks remained. Data on average land values were obtained from the 1950 Census of Housing, and block values were imputed to the households concerned. This latter procedure, while somewhat questionable, seems justified in the light of the study by Hoyt (1939).

The over-the-road distance of each household in the sample from the nearest

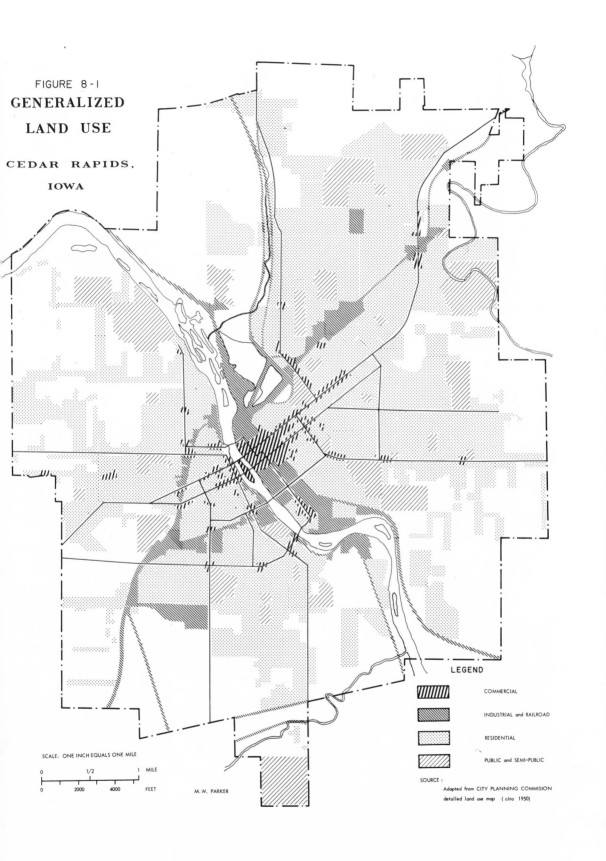

FIGURE 8-1
GENERALIZED
LAND USE

CEDAR RAPIDS,
IOWA

SCALE: ONE INCH EQUALS ONE MILE

0   1/2   1   MILE

0   2000   4000   FEET

M. W. PARKER

LEGEND

COMMERCIAL

INDUSTRIAL and RAILROAD

RESIDENTIAL

PUBLIC and SEMI-PUBLIC

SOURCE :

Adapted from CITY PLANNING COMMISION
detailed land use map   (ciro 1950)

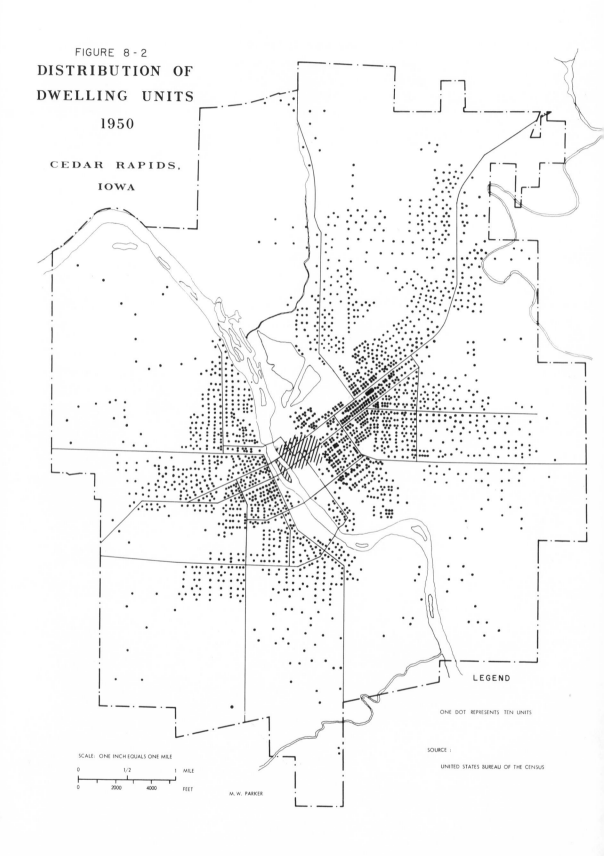

FIGURE 8-2

DISTRIBUTION OF
DWELLING UNITS
1950

CEDAR RAPIDS,
IOWA

LEGEND

ONE DOT REPRESENTS TEN UNITS

SOURCE :

UNITED STATES BUREAU OF THE CENSUS

SCALE: ONE INCH EQUALS ONE MILE

| 0 | 1/2 | 1 | MILE |

| 0 | 2000 | 4000 | FEET |

M. W. PARKER

member of each of the four types of commercial nuclei observed in the city was obtained. Because of an observed high correlation between distance to the CBD and distance to the nearest high order center, the latter measure was dropped from the model and the distance to the place of work of the head of the household added. The model fitted to these data was the logarithmic form discussed previously.

The Value Study

Standard multiple regression techniques were again utilized to examine the associations of the parameters of the model. The following estimating equation resulted:

$$Y = \frac{1000\ X_1{}^{.050}X_2{}^{.087}X_3{}^{.024}}{X_4{}^{.039}}$$

where:

X_1 = the road distance in miles from the CBD
X_2 = the road distance in miles from the nearest medium level center
X_3 = the road distance in miles from the nearest low order center
X_4 = the road distance in miles from the place of work of the head of the household
Y = average value of single family dwelling units in the block, in tens of dollars

None of the coefficients was significant at the 5 per cent level, and only one, X_2, was significant at the 20 per cent level. In light of this fact it is not surprising that the model explained less than 10 per cent of the observed variation in land values (R^2 = 0.07).

The Rent Study

Mean rental values as well as average land values were reported for 51 blocks of the 71 block value sample. The rental data represent, in terms of the way in which the information was obtained, a somewhat more accurate measure than the value data. The same four parameter model used in the value study was fitted to the block rental data, and the following estimating equation obtained:

$$Y = \frac{44\ X_1{}^{.131}X_2{}^{.053}}{X_3{}^{.006}X_4{}^{.044}}$$

Once again none of the coefficients was significant at the 5 per cent level, and only one, X_1, was significant at the 20 per cent level. Slightly over 10 per cent of the observed variation was accounted for by concomitant variation in the postulated independent variables (R^2 = 0.12).

The unimpressive performance of the nonlinear model in these two cases seems to indicate that neither the form of the model nor the way in which the distances were weighted represents the actual fashion in which location contributes to the value of residential sites. In his discussion of the work done in Washington, Garrison (1957) pointed out that " . . . simple linear measurements of roads were unusable . . . successful distance measurements were weighted in terms of road type and use (purpose and amount). This implies that simple linear concepts of space are not very useful. " The current studies tend to confirm this point, and to indicate that simple weighting schemes (e. g. , the logarithmic weighting of the last model) do not increase the usefulness of simple linear measurements of spatial relationships.

Transport Inputs and the Value of Residential Sites

The study in Washington State by Garrison (1956) utilized a linear model where the measures of residential location were in terms of miles traveled per year for a specific trip purpose over a given road type. The result was an estimate of the contribution of each trip type, and of each road type, to the value of the property. The measures utilized may be regarded as simple linear distances weighted by trip frequencies or, more in line with current location theory, as several disaggregated measures of the transport inputs to the households occupying the sites.

A discussion of the theoretical structure which relates transport inputs to the value of residential sites will be deferred until later. However, certain of the basic assumptions involved in the use of simple models, such as the ones utilized in the Washington State studies, may be commented upon. The ability to use such simple models (the term *simple* is used here to denote linear, single equation forms) is dependent upon the correctness of two assumptions: (1) that trip frequencies do not vary significantly with distance, and (2) that the demand for movement is independent of the quality of road service available. Examination of the data collected in the Washington State studies failed to reveal any reason to believe that these assumptions were invalid for specific trip types even though current interaction constructs (see Iklé, 1954) and common knowledge would seem to hold otherwise.

The study of the structure of transport inputs to individual households has received little attention, either from the theoretical or the empirical point of view. In light of this lack of information, and the somewhat puzzling empirical results reported by Garrison (1956), a careful examination of the topic would seem to be indicated before proceeding further with the current discussion. The following chapter reviews the currently available theoretical structures pertaining to the demand for transportation at residential sites as well as the small number of empirical studies which have been conducted on this topic. The results of several studies undertaken utilizing the detailed travel data from Cedar Rapids, Iowa, are also presented.

9

The Demand for Transportation
at Urban Residential Sites

THE STRUCTURE of the demand for movement by individual consumers or households has received little attention in the past. Theoretical constructs and empirical
studies have, for the most part, been concerned with broad patterns of movement
as exhibited by large masses of people while the study of the spatial behavior patterns of individual consumers and households has been neglected, as is evident
from the small number of theoretical works and good empirical studies available
on this topic.

The two previous chapters in this section attempted to explore some of the relationships postulated between location and the value of residential sites. As a result of this exploration, it became apparent that a higher level of information on
transport inputs to individual households was necessary before adequate explanations of the way in which value varied with location could be put forth. This chapter
examines the current level of theoretical and empirical knowledge and attempts to
extend this level through several empirical studies of the factors associated with
transport inputs to individual households.

Previous Theoretical Formulations

In order to review the previous theoretical structures relating to the demand for
transportation at urban residential sites it is necessary to turn to such fields as
location theory, transport economics, and marketing research to obtain even fragmentary statements pertaining to the topic under review.

General Location Theory

Researchers working in the field of general location theory have given only a
small amount of attention to the spatial problems of the individual consumer. Lefeber (1958) does not even discuss optimal conditions for consumers, and Isard
(1956) admits that current concepts in location theory are unable to account for
variations in transport inputs from household to household.

Isard does, however, provide some indication that spatial relationships may not
be the only factors of importance when he introduces the concept of the consumer's
space preference (which is defined as a measure of the individual's desired level
of social contact). Different individuals placed in the same spatial situation with
identical levels of information may behave differently, and it is postulated that
these differences in behavior arise out of the differences in the space preferences

157

of the individuals involved. These space preferences (which we might consider to be expressed as a set of desired trip frequencies) are held to be determined by social and psychological forces which are exogenous to the spatial system.

In addition to the work of the location theorists, a little known analysis by Troxel of the demand for the movement of persons provides some additional insights into the decision-making process of the individual consumer.

Troxel's Theory of Demand

One major discussion of the theory of consumer demand for movement is that presented by Troxel (1955). His analysis deals only with certain movement types, specifically, regular patterns of person movements as defined on the basis of the individual person or household. Among the movement types excluded from the discussion are movements which are directly associated with the movement of goods as well as movements of persons for governmental and military purposes. Basic to Troxel's theoretical discussion are four major concepts pertaining to the movement of persons: (1) the home base, (2) the round trip unit of travel, (3) differentiation of travel purposes, and (4) recurring patterns of movement.

The demand for any movement in space may be viewed as having its origin in a point or area limited in space. Many demand points exist either potentially or actually in any sizable land area, and such conventional identifications as home, farm, factory, etc. have been given to them. Home locations are common organization points in family and community life, and they are frequent reference points in the actual behavior of travelers. Thus the concept of the home base as a starting and ending point for most journeys may be recognized.

Given the concept of the home as a demand point in space, the travel of persons is expressed primarily in terms of round trip movements to and from this point. A person or family has a desire to go out from a home location and to return to it. In the organization of person movements adopted here, the home is recognized as one end point and the other end point (or set of end points) is located somewhere distant in space. Thus most person trips may be considered in light of the concept of the round trip unit of travel.

The trip end points beyond the home afford a variety of products and services, and persons show a willingness to move different distances and to incur different time outlays for different end results. This has been observed empirically many times (for instance, see Table 9-I). Thus, there seems to be a definite differentiation of travel purposes in terms of both frequency of occurrence and distance traveled.

Movement through space requires the expenditure of some amount of time by the traveler, an amount of time that varies with the distance traveled and the means of the travel. Movement occurrences, it is felt, tend to be expressed repetitiously through successive time periods. For instance, the week period seems to be a prominent one in family travel that is routinized to a common and recurrent form from time period to time period. This has been noted empirically by both the Traffic Audit Bureau and Mills and Rockleys (1955).

The sources of the demand relation for movement in space are the products of being at home and not at home--the differences between what may be obtained away from home and what must be foregone at home and vice versa, since a person or

TABLE 9-I
TRIP PURPOSES AND DISTANCES IN FOUR STATES*

Purpose	Average Trip Length (Miles)	Vehicle Miles (Per Cent)	Number of Trips (Per Cent)
Shopping	3.7	6.0	13.4
Educational, civic, religious	5.0	2.8	4.6
To and from work	5.9	22.7	32.2
Business and farming	10.4	21.3	17.0
Pleasure riding	13.0	14.1	9.0
Medical and dental	15.5	2.5	1.3
Vacation	249.7	4.8	0.2
Other purposes		25.8	22.3

*Adapted from Troxel (1955) and Bostic *et al.* (1954).

family unit foregoes something that is obtainable at home during the amount of time used traveling away from the home location. In the following discussion it is assumed that locations of homes are given; no distinctions are made between transport techniques, and no specific time period is designated although it may intuitively be considered to be a week.

The total anticipated product of travel for a person or family unit commonly varies with the total time that the person or family is absent from the home location. The total product of travel--the income from more hours of work, shopping bargains from more searching around, and the many other factors--increases as time away from home increases. But the total travel product, as suggested in Figure 9-I, presumably increases more slowly after some limit of elapsed time is reached. This decreasing amount of increase may be inferred in part from differences in travel purposes. In Figure 9-I the order is from the largest to the least travel product per successive time unit, i.e., a maximum travel product is obtained at any limit of elapsed time. Taking the most productive purposes first and distributing his time and movement in some order of product expectation, the traveler comes to lesser purposes in the later time units. Decreasing amounts of total products may be inferred, too, from the probability of less satisfaction from successive time units that are devoted to the same travel purpose. As also illustrated in Figure 9-1, the speed of available movement affects the anticipated travel products in relation to travel time. A higher speed allows the traveler to achieve more purposes or to obtain more products within any trip purpose in a given time period.

Implicit in the foregoing discussion is the idea of deductions from the total travel product. Traveling is not all profit as most of us know. The traveler must deduct for losses of products or satisfactions which would be generated by remaining at home or which are foregone by his not traveling to other locations. This loss curve is displayed in Figure 9-2. In consequence of the rational behavior of the traveler, the curve of the losses in total satisfaction shows increasing amounts of increase

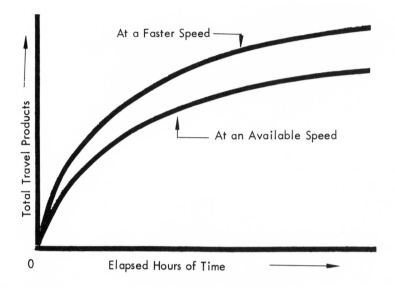

Figure 9-1   Total Travel Products

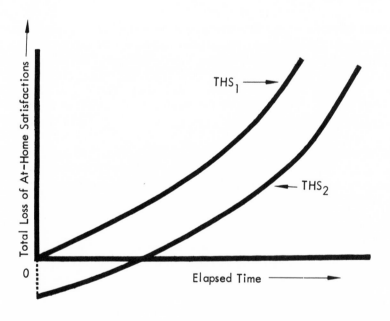

Figure 9-2   Total Travel Losses

as the elapsed time increases (it may also be recognized that to some persons there is a net loss in being too much at home--this is illustrated by the negative value of the first portion of the THS_2 curve in Figure 9-2). Other deductions from the total travel product must be made owing to the effort involved in movement-- the so-called friction of distance.

Having established an aggregate travel product in an elapsed time relation and a function of deductions from the total travel product, it is now possible to combine them to produce a net product schedule. This net travel products schedule (Figure 9-3) indicates that the traveler comes to an amount of elapsed time when further additions in travel involve decrements in net products; these decrements can increase as more time passes. From the net travel product schedule, the schedule of marginal net products (first derivative of NTP) is obtained. This marginal product curve (Figure 9-4) slopes downward, in further time extensions becomes zero, and eventually passes on to a minus value. This action on its part is indicative of the downward turn of NTP schedule after reaching some maximal point (t_1). The zero limit of the MNP schedule is the equivalent of a zero price limit, i.e., beyond time t_1 the traveler must be paid to travel further. The curve of the MNP schedule conforms in its general shape to a travel demand schedule. If the product values are turned into prices of payment, the MNP series becomes a demand price schedule in units of elapsed time.

Referring again to Figure 9-1, it may be seen that the marginal net product increases as the available speed of movement increases. Increases in travel demands can be expected as more speed, more safety, and so forth, affect access to more net products per time period. The schedule of marginal net products is also a function of the spatial limits within which alternative movements can be chosen. Restrict the limits of possible travel and the marginal net product curve decreases, i.e., if the maximum radial distance beyond the home position is reduced successfully, then the marginal net product values (or travel demand) in units of elapsed time can be expected to decrease too. This effect follows because, in lesser spatial limits, the access to net products (or range of alternative choices) is reduced.

The total net products of travel may be differentiated among various transport situations. For instance, there are possible differences between individuals, age groups, and so forth. Variations between purposes are also evident. Additional time is not used to one end if, as an expectation, it promises a larger addition to the net product in some other use. Thus, in terms of the elapsed time where the marginal net profit equals zero, purposes are not expected to come out as equivalents. More adult time away from home goes commonly into employment, visiting, or recreation than into medical care, luxury, or specialty shopping (see Figure 9-5).

The elapsed time and the distance traveled are related. In the foregoing discussion, the term *elapsed time* has been used to refer to time spent away from home; thus only part of the time is used in actual movement. Referring to a single trip or type of trip, the distance relation in the net total and marginal products of movement can be expressed in successively longer trip lengths beyond the home base, or, when standard distance units such as miles are used, the demand relationships may refer to the number as well as the lengths of trips.

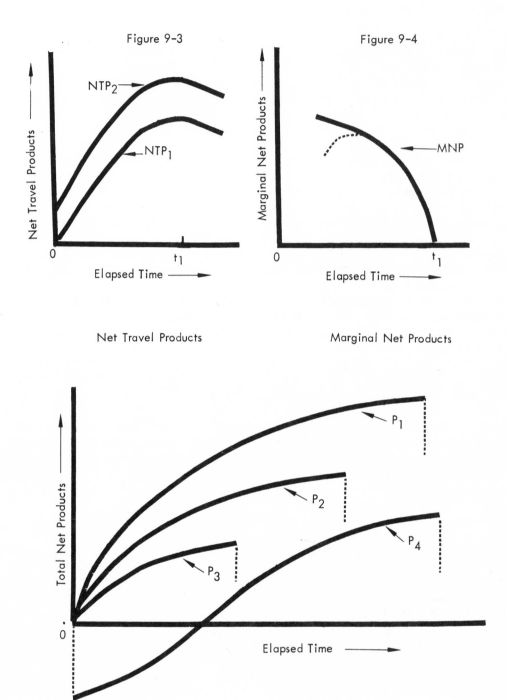

Figure 9-3

Figure 9-4

Net Travel Products

Marginal Net Products

Figure 9-5

Net Products of Various Trip Types

A distance series of marginal net products may be conceived of for a single trip for a specific purpose. The traveler can project trip paths in several directions or in a roundabout fashion beyond the home. Any such path involves a discrete series of point locations, and any traveler may have many of these trips paths that contain sequential expectations of marginal net products. As many paths exist as travel purposes, combinations of purposes, and possible path locations--a very large but still finite number.

In the preceding discussion the location of the home base has been assumed to be given and fixed for each individual or family unit, and thus relations that were significantly reflected in the choice of home sites have been necessarily excluded. However, the decision limits of distance are readily noted where, for a given trip purpose, the distance that must be traveled in order to attain this purpose is so great that the home base location must be changed. A simple example here would be a person whose job site has been shifted from one part of the city to another and who finds the new distance relationship between home and work to be such that it is no longer practicable to utilize the old home base.

This detailed discussion of the theoretical structure formulated by Troxel may be justified entirely on the basis of the theory's content. However, it is interesting to note that it currently represents the only general theory, aside from portions of location theory, which deals with the problem of the movements of individual consumers in space. While it is unfortunate for our purpose that elapsed time rather than distance was utilized as an organizing factor, the impact of highway improvements upon demand is still evident.

The Demand for Movement as a Function of Traffic Conditions

In Troxel's general statement, problems of cost and difficulty of movement were considered only implicitly as part of the total travel losses schedule (Figure 9-2). Recently, Beckman *et al.* (1956) have formulated a model of short-run influences on (1) the choice of individual free speeds, (2) the selection of routes, and (3) the demand for road transportation. The discussion here will be directed to the latter topic. Rather than enter into a detailed discussion of the models, only indications of their magnitude and certain general findings of the author's work will be pointed out.

Beckman *et al.* assume the demand for transportation between a given origin and destination to be a function of the average trip cost between these locations. The curve shown in Figure 9-6 represents the number of trips per given time period which are undertaken at various average cost levels. This is, of course, the general demand curve. In certain circumstances, when the number of trips is independent of traffic conditions, the curve is vertical, and demand can then be described completely by fixed origin-destination figures. The demand curve also indicates the total benefits derived by the population of road users at a given level of average trip cost (the shaded area between the demand curve and the average cost line).

Given this notion of demand as a function of average trip cost, the authors examine the factors influencing the average cost level and find them to be flow and capacity, variables which are in turn dependent upon average trip cost levels. The authors then undertake to show that there exists one level of flow in which traffic conditions give rise to a demand which just equals the prevailing flow. This

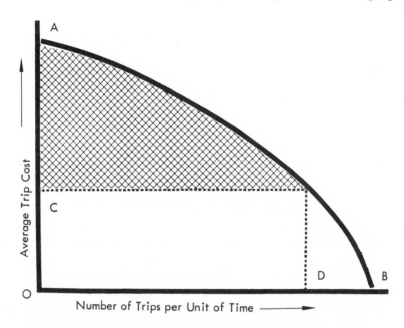

Figure 9-6   Demand Curve for Travel Between Two Locations

is defined as the equilibrium state. Given proof of the existence of such an equilibrium, the question may then be raised as to the effect of changes in capacity and demand upon the system. Such changes will destroy the existing equilibrium situation and will lead to some new and different situation. The authors show that this new situation will be a stable equilibrium.

This brief discussion has failed to bring out more than a few of the more obvious points in the authors' discussion. For more detailed statements regarding this short-run model the reader is referred to the original work.

A Simple Shopping Model

A recent report by the firm of Alderson and Sessions (1957) on some basic facets of consumer behavior included a theoretical model of consumer and retailer behavior in terms of variety and distance relationships. The form of this model was discussed earlier (see Chapter 3), so only those portions having a direct bearing on the demand for movement will be discussed here.

It is assumed that a shopper does not know (with certainty) in advance whether he will obtain what he wants by shopping at a given store. The greater the number of items carried by the store, the greater ordinarily is the customer's expectation of success. Assign to this expectation a function denoted by $p(N)$ which varies between limits of 0 and 1, with 1 representing certain foreknowledge of success and 0 representing certain foreknowledge of failure.

In going to and shopping at some particular store, the customer incurs certain costs. If the distance of the customer from the store is D, it is assumed that for him the costs of movement are strictly proportional to D, and are given by C_dD where C_d is some constant.

The model also includes total opportunity cost, i.e., those costs incurred by not shopping elsewhere and held to be independent of the number of items sold and the distance of the consumer's home base from the store. The total opportunity cost is denoted in the model by C_i.

Total shopping costs are assumed to be the sum of the distance cost, the total opportunity costs, and a term representing a store size cost, i.e.,

$$C_n D + C_n \sqrt{N} + C_i$$

This sum represents the total decrement to the total travel product in shopping. This is a more explicit formulation than Troxel presented and one in which movement costs are included; however, the aggregation of opportunity costs and the assumptions pertaining to movement costs leave a great deal to be desired.

Certain economic implications may be drawn from the model; among those of interest in the present context are (1) that the minimum number of items necessary to induce a customer to shop at a given store will increase with D, that is, the high shopping costs of the distant customer can only be overcome by a high probability of a successful shopping trip, and concurrently (2) that for every value of N (the number of items carried) there will be a maximum consumer distance from this store. Thus the net benefit to a customer shopping at a given store is a function of the locational choice made by the consumer in selecting his residence.

Demographic Variation and Consumer Behavior

Economists have occasionally examined the impact of demographic variation upon a series of economic variables. The study of Duesenberry (1949) may be cited as an example of a theoretical formulation which examined the impact of such factors as age and previous economic status upon expenditure behavior of households. Empirical studies have been carried out by a number of workers, and the recent works by Klein (1954) and Zwick (1957) provide examples of the way in which these variables may be entered into empirical studies.

The only theoretical formulation available which relates demographic variations to possible variations in consumer behavior in space is that of Isard (1956). Demographic variations would be associated with variations in the space preferences of individual consumers (the concept of space preference was introduced earlier in this chapter) and, hence, with variations in spatial behavior patterns.

Previous Empirical Studies

The continuing interest in traffic forecasts by highway planners has brought about many empirical studies of the movement of persons within the urban areas of the United States. The majority of the studies have been of the origin-destination type and have been confined to obtaining cross-section views of the zone-to-zone movements of persons within an urban area. A summary of the standard techniques utilized in these studies as well as a listing of the majority of the studies which have been completed may be found in the review by Barkley (1951). A recent article by Curtiss (1957) points out several ways in which these studies have failed to provide either the desired estimates of future movements or even accurate descriptions of present movement patterns. From the large number of studies which have concerned themselves with the movement of persons in general, it is possible to select a much smaller number which have dealt with the travel patterns of individual households on a more specific basis. The Washington State studies would be included in this latter category and their results have been discussed earlier in this section (see chapter 8).

Gardner's Study of Baltimore

Gardner (1949) examined twelve study areas within the city of Baltimore in the hope of discovering associations between gross patterns of movement and the socioeconomic characteristics of the areas. The areas were selected to represent six socioeconomic types ranging from blighted areas near the central business district to outlying areas of high residential quality. Traffic flows were tabulated by areas in terms of their distribution by purpose and mode as well as by total volume. The traffic flow figures were then compared with the socioeconomic characteristics of the areas to determine what associations existed.

While this study suffered from a lack of powerful analytic tools, the author did conclude that (1) the number of work trips per resident was fairly constant from neighborhood to neighborhood but the number of trips made for other purposes tended to increase in proportion to the wealth of the neighborhood, (2) population characteristics such as age and occupational structure tend to have quite noticeable effects on travel patterns, and (3) the total number of trips per resident increased as distance from the central city increased. The latter characterisitc seemed to be relatively independent of the economic classification of the areas involved.

The Virginia Studies

In 1951 the University of Virginia in cooperation with the Virginia Department of Highways and the Bureau of Public Roads undertook to examine the relationships existing between socioeconomic characteristics and travel habits of the people in Charlotte County, Virginia. Upon the basis of the detailed data collected it was concluded that (1) both trip frequency and trip mileage were related to the family income level in a positive fashion, (2) as the socioeconomic status increases the average daily trips and miles rise, and (3) the amount of travel seems to be directly related to the composition of the family group.

It was pointed out earlier that no theoretical formulations are currently available which relate demographic variations to consumer behavior in space. The results of the Virginia studies seem to indicate that these factors do indeed exert a considerable influence upon household travel patterns, and the recent study on family pleasure travel by O'Brien (1958) also tends to confirm this observation.

Hamburg's Study of Detroit

Utilizing information collected by the Detroit Metropolitan Area Traffic Survey, Hamburg (1957) conducted a study which examined the relationship of vehicular trip frequency per dwelling place to several factors. These included: airline distance from the central business district (weighted logarithmically), household income, family size, automobile ownership, and residential density.

All information was imputed to households from subzone data. For instance, trips per dwelling place were obtained by dividing the total trips of a small area by the number of dwelling units in the area. Values of other variables were obtained in a similar fashion.

Multiple regression analysis was used to examine the associations between the variables. The author found trip frequency to vary in a systematic manner with the measures of residence location, family size, and automobile ownership. The measures of residential density and family income were not significantly associated with trip frequency. It should be noted that while this study shows a significant degree of association between *vehicular* trip frequency and residential location, no statements regarding the relationship of over-all trip frequency (which includes walking trips) to residential location may be developed from its results.

Mertz and Hamner's Study of Washington, D. C.

Mertz and Hamner (1957) conducted a study in Washington, D. C., which was quite similar to the one undertaken in Detroit by Hamburg. Multiple regression analysis was again used to examine the associations between vehicular trips per dwelling unit, automobile ownership, population density, household income, and distance (presumably airline distance) from the central business district. Household characteristics were imputed from census tract data in a manner similar to that employed by Hamburg in Detroit.

The distance and income factors did not prove to be statistically significant, and population density was of doubtful value. The measure of automobile ownership proved to be highly significant, and it was concluded that little additional accuracy could be obtained by adding other factors. Linear models were utilized throughout this study, although an examination of the scatter diagrams suggests the possibility of nonlinear relationships in the distance and income factors.

TAB's Studies in Fort Wayne and Cedar Rapids

During the years immediately following World War II the Traffic Audit Bureau, Inc., conducted a series of studies designed to devise new methods for the evaluation of outdoor poster advertising. The first of these studies was undertaken in the Fort Wayne Standard Metropolitan Area (SMA) in 1946 and examined, among other things, the travel patterns of a sample of the residents of Allen County. The information collected pertained to all trips made in one day and included data on routes traveled, mode of travel, and trip purposes. A certain amount of socioeconomic information on the households in the sample was also collected.

While the Fort Wayne study provided adequate information on the coverage attained by poster showings, additional information on repetition was needed to assess the impact of different showings. To obtain this information, a second study was undertaken in Cedar Rapids where informants were paid to keep detailed travel diaries for a thirty-day period. Fairly detailed socioeconomic information on the households involved was obtained at the same time. While the conclusions reached as a result of these studies are of little direct interest here, the Traffic Audit Bureau and the Outdoor Advertising Association of America have made the basic data available for further analysis.

Studies in Other Countries

At present, little information is at hand relating to studies of passenger movement in the urban areas of other nations. Since the present preliminary study is mainly concerned with intraurban rather than interurban movements of persons, the work by Godlund (1956a, 1956b) and Kant (1957) in Sweden, Dickinson (1957) in the Netherlands and Belgium, and other similar works elsewhere will not be discussed here. However, in 1954 a study was undertaken in nine English towns by the firm of Mills and Rockleys, Ltd., which in its scope and nature was quite similar to that undertaken in Cedar Rapids by the Traffic Audit Bureau. A random sample of households was selected in each of the nine towns, and each household was interviewed to determine the nature of its travel during the last seven days. The questions were specifically directed toward determining the number of passages past certain preselected points within the cities, and while information on exact routes and trip purposes was obtained during the interviews, it was not recorded. While this study was directed toward problems of audience coverage and impact of outdoor poster advertising, Mills and Rockleys did note that the movement of the persons within the sample tended to follow a seven day cycle and that the routes followed and modes of travel tended to be highly repetitious.

While little is known at present about urban passenger movements in the Soviet Union, a recent study of Soviet transportation by Hunter (1957) does present a few comments based on currently available information.

The foregoing review of existing theory and practice leaves a somewhat confused picture as to both the theoretical and actual relationships between trip frequency, transport inputs, and location of the consumer's residence. On the basis of information made available by the Traffic Audit Bureau, Inc., and the Outdoor Advertising Association of America, several empirical studies will be attempted to try to extend our knowledge of the situation.

Survey Methods in Cedar Rapids and Linn County

The following discussion of the survey methods used in the Cedar Rapids movement study is drawn largely from unpublished materials furnished by Mr. Victor Pelz of the Traffic Audit Bureau. The discussion is pertinent to the analysis to follow in this chapter and to the analysis in Chapter 11 where the Cedar Rapids data are also used. Chapter 11 contains an analysis indicating that Cedar Rapids is typical of urban centers of its size class in important aspects of land use and traffic.

Sample Design

A minimum sample of 256 dwelling units was established for the study on the basis of this figure being close to 1 per cent of the occupied dwelling units in Linn County in 1940. This minimum total of 256 households was then allocated to Cedar Rapids proper, rural farm, and rural nonfarm (including the town of Marion) on the basis of the estimated number of dwelling units. The sample was based on R. L. Polk Co., City Directory for Cedar Rapids, dated 1947. This information was updated with additional information provided by the city engineer to include new dwellings constructed up to April 1, 1949.

Since this survey required informants (including every member of the household ten years of age and over) to keep records of all travel (by all modes, including trips on foot) for 30 consecutive days, it was necessary to provide for a sizable shrinkage caused by families who would start to keep records but drop out before the 30 days were over. This problem was met by starting with a sample of 308 instead of 256. It was also recognized that many families would refuse even to start a 30-day consecutive record. This problem was met by drawing two samples of 308 each and pairing on a map each family in the first sample with the nearest family in the second sample. A family refusing in the first sample was replaced by the paired family in the second sample. The rate of refusals in the first sample was approximately 16.6 per cent, but in no case did the second family of a pair refuse. Actually there were 262 households in which all members completed the full 30-day record. Since the total of 262 households was not distributed properly in the three categories (Cedar Rapids, rural farm, and rural nonfarm), six families were selected at random to be dropped from the final sample.

In selecting dwelling units to be included in the sample, every dwelling unit was given an equal chance of coming to the sample. The household selected from the list of addresses was chosen by taking every address after a starting point had been selected from a table of random numbers. The same procedure was followed for selecting dwelling units in the areas of new construction. In both cases the address was used as the sampling unit even though the city directory furnished a name as well as the address. Similar methods were used to select dwelling units located outside the corporate limits of Cedar Rapids.

Interviewing

Street addresses of households for the interviews in Cedar Rapids were assigned to each interviewer. In the event that the household from the first sample refused to participate, the interviewer then approached paired households from the second sample. Interviewers were instructed to make whatever call-backs were necessary to contact the households in the first sample before approaching the second sample household.

The purpose of the survey was explained as one designed to secure information over a consecutive 30-day period as to where people travel. Interviewers were instructed to answer questions of respondents fully and freely. While the fact that the survey was being done on behalf of outdoor advertising was not volunteered by interviewers, this explanation was freely made if needed to satisfy respondents.

Responsibility for seeing that daily travel diaries were made out properly was placed either on each individual respondent or upon some one person in the family, depending on which method seemed to the interviewer most likely to be effective.

Each household was paid a maximum fifteen dollars for acceptable completion of 30-day diaries from each member of the household, 10 years of age and over. The total interviewing period covered five weeks. A payment of two and a half dollars was made for each of the four weeks, plus a premium of five dollars to those families who completed the full 30 days. Families dropping out before the end of the 30 days were paid pro rata for the time for which they actually completed diaries.

After the initial interview, the interviewer returned to the home to check up on diaries made out by those not at home at the time of the initial interview and to inspect diaries for the preceding day. If all respondents in the home then seemed to understand the making out of the diaries, they were accepted; otherwise further instructions were given. From then on the interviewer called at each home every other day to review diaries and to pick them up and bring them into the office. This every-other-day contact was continued throughout the 30 days.

Completed diaries brought in by interviewers were checked day by day against a record kept for each individual. Each day's diaries were reviewed and, if not clear or satisfactory, were referred back to the interviewer to be taken up with the respondent on the next visit.

Modifications of the Sample

Because of time and resource limitations, it was impossible to utilize in the current preliminary studies all the information collected by TAB. The empirical discussions which follow in this chapter, as well as the discussions in Section IV, are based upon a subsample of 100 households located within the corporate limits of the city of Cedar Rapids (see Figure 9-7). Detailed tabulations were undertaken on the movements of these households during the second and third week of the sample period. Extensions of these tabulations to all households and the full time period is currently underway.

Transport Inputs to Households in Cedar Rapids

Tabulation of the movement data collected in Cedar Rapids revealed a very complex pattern of movement. This complexity was especially evident in the structuring of the movement of persons by purpose of movement. In many surveys, movement purposes are classified under several general headings (e.g., those used in Table 9-I). It proved to be impossible to use these simple structurings in classifying the Cedar Rapids data, and even further disaggregation failed to solve the problem. For instance, during the study period some 144 stops were made at supermarkets by individuals in the travel sample. While 21 per cent of these stops did not involve other movement purposes, 34 per cent involved one or more stops for other shopping purposes, 19 per cent involved stops at the place of work, and 15 per cent involved stops for other purposes such as visiting friends (see Table 9-II and the more detailed discussion in Chapter 11).

Because of this complexity the preliminary studies reported here were forced to deal with gross measures of the demand for movement rather than with more disaggregative measures. The gross measures utilized here are: (1) total time away from home per household per time period, (2) total trips per household per time period, and (3) total distance traveled per household per time period. These gross measurements specifically exclude movements made for the sole purpose of visiting friends or for the pleasure driving.

The General Form of the Model

The review of previous theoretical and empirical work suggests three sets of factors which may be postulated to influence the demand for transportation and the location of the residence. These factors are entered into a general linear model which was applied to each of the three gross measures of the demand for movement. This model takes the form of the first model discussed in Chapter 8, viz.,

$$Y = \beta_1 x_1 + \beta_2 x_2 t \ldots t \beta_n x_n$$

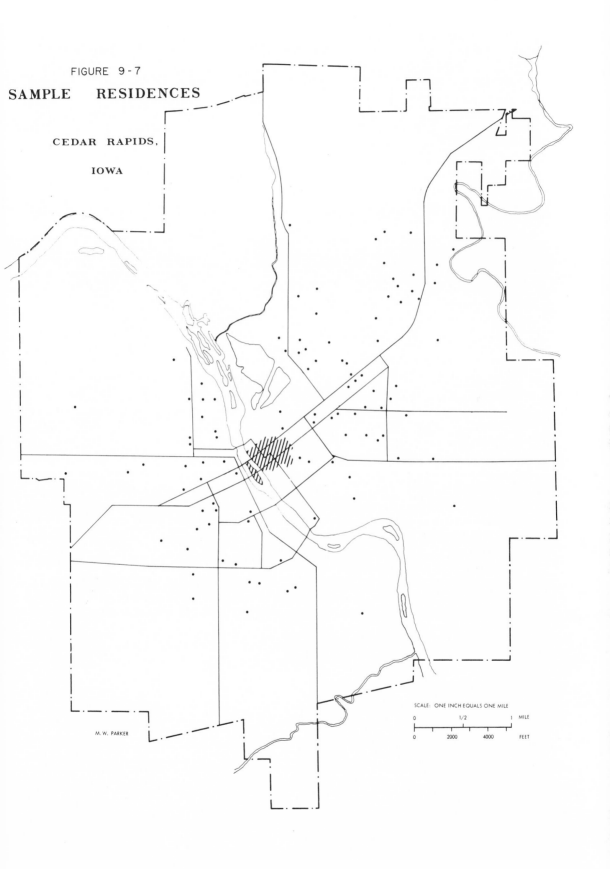

FIGURE 9-7

SAMPLE RESIDENCES

CEDAR RAPIDS,

IOWA

M. W. PARKER

SCALE: ONE INCH EQUALS ONE MILE

| 0 | 1/2 | 1 | MILE |

| 0 | 2000 | 4000 | FEET |

TABLE 9-II
STOPS FOR SHOPPING; BY FUNCTIONS; ON TRIPS DURING A TWO-WEEK
PERIOD BY SAMPLE HOUSEHOLDS IN CEDAR RAPIDS, IOWA

Function	Number of Stops	Per Cent of Total	(1)	(2)	(3)	(4)
Grocery	256	15.0	61.3	12.9	9.4	16.4
Drug store	158	9.3	38.6	30.4	15.2	15.8
Department store	155	9.1	11.6	56.1	25.2	7.1
Supermarket	144	8.5	21.2	34.0	19.4	15.3
Restaurant	142	8.4	9.2	19.7	59.8	11.3
Theater & other amusement	107	6.3	64.5	20.6	5.6	9.3
Variety	63	3.7	4.8	58.7	31.7	4.8
Gas station	62	3.6	21.0	14.5	40.3	24.2
Bank	61	3.6	8.2	29.5	52.4	9.8
Clothing & shoe	50	2.9	6.0	66.0	18.0	10.0
Doctor	50	2.9	38.0	42.0	14.0	6.0
Misc. retail	38	2.2	10.5	47.4	23.7	18.4
Bakery	33	1.9	15.2	48.5	24.2	12.1
Auto repair	29	1.7	24.1	27.6	34.5	13.8
Confectionary & ice cream	29	1.7	44.8	20.7	. . .	34.5
Hardware	26	1.5	23.1	42.3	15.4	19.2
Meat-fish & fruit-veg. market	20	1.2	20.0	55.0	10.0	15.0
Business offices (to pay bills)	19	1.1	26.3	52.6	15.8	5.3
Barber	17	1.0	29.4	35.3	29.4	5.9
Real estate, loans & insurance	17	1.0	11.8	70.6	11.8	5.9
Furniture	17	1.0	17.6	58.8	17.6	5.9
Auto accessory	16	0.9	31.2	18.8	31.2	18.8
Tavern	15	0.9	13.3	20.0	60.0	6.7
Beauty shop	14	0.8	28.6	35.7	21.4	14.3
Cleaner & laundry	13	0.8	. . .	46.2	38.5	15.4
Motel-hotel	12	0.7	25.0	33.3	16.7	25.0
Appliance	11	0.6	18.2	45.4	18.2	18.2
Dentist	10	0.6	60.0	20.0	20.0	. . .
Shoe repair	9	*	. . .	66.7	33.3	. . .
Farm equipment & supplies	9	*	11.1	44.4	22.2	22.2
Auto dealer (new)	8	*	. . .	75.0	12.5	12.5
Radio-TV, sales & service	5	*	. . .	80.0	20.0	. . .
Funeral home	5	*	40.0	40.0	. . .	20.0
Auto dealer (used)	5	*	60.0	20.0	20.0	. . .
Other building supplies	4	*	. . .	75.0	25.0	. . .
Other professional	4	*	50.0	50.0	. . .	. . .
Lumber yard	3	*	. . .	33.3	33.3	33.3
Coal-fuel dealer	3	*	. . .	100.0	. . .	. . .
Fix-it, bicycle & welding	2	*	. . .	. . .	100.0	. . .
Building service	0	*	. . .	. . .	. . .	. . .
Furniture repair	0	*	. . .	. . .	. . .	. . .
Unspecified	59	3.5	5.1	69.5	6.8	18.6

(1) Per cent single-purpose stops.
(2) Per cent of stops associated with one or more other shopping stops.
(3) Per cent of stops associated with stop at place of work.
(4) Per cent of stops associated with one or more stops for other purposes
(e.g., visiting friends).
*Less than 0.5 per cent.

In this case the variables will be:

Y = some measure of transport demand by the household

Measures of socioeconomic factors:

x_1 = size of family

x_2 = age of the head of the households in years

x_3 = has the head of the household received any education beyond the high school level? (1 = yes and 0 = no)

x_4 = is the household in the low income (below \$3,000 per year) group? (1 = yes and 0 = no)

x_5 = is the household in the medium income (\$3,000 to \$5,000 per year) group? (1 = yes and 0 = no)

x_6 = is the household in the high income (above \$5,000 per year) group? (1 = yes and 0 = no)

x_7 = are there one or more workers in the household? (0 = yes and 1 = no)

x_8 = are there one or more school children in the household? (1 = yes and 0 = no)

Measures of the availability of transport:

x_9 = does the family own an automobile? (1 = yes and 0 = no)

x_{10} = road distance in miles to the nearest transit line

Measures of residence location:

x_{11} = road distance in miles to the central business district

x_{12} = road distance in miles to the nearest high order retail center

x_{13} = road distance in miles to the nearest medium order retail center

x_{14} = road distance in miles to the nearest low order retail center

Of the 100 households in the travel sample, 11 were dropped for various reasons (including poor reporting and inclusion of business-related travel). Thus the results of the following empirical studies are based upon information obtained from 89 households. Summary measures of household characteristics are tabulated in Table 9-III, and some simple measures of association are shown in Table 9-IV.

The Elapsed Time Study

Troxel (1955) has suggested the use of the amount of time spent away from home as a measure of the demand for movement at residential sites. The Cedar Rapids travel diaries contained information on the time at which each trip began and ended. The duration of each trip was obtained from these entries and the information aggregated by households

Preliminary computations disclosed that members of the average household spent about 164 hours away from home during the study period. This amounted to some 28 hours per week per person. An examination of the zero order correlations (see Table 9-IV) discloses significant associations only with size of family, a socioeconomic variable, and total number of trips, another gross measure of demand.

Fitting the general linear model to the household data produced the following estimating equation:

TABLE 9-III
AVERAGE HOUSEHOLD CHARACTERISTICS
CEDAR RAPIDS TRAVEL SAMPLE

Mean size of family	3.1	persons
Mean age of head of household	44.6	years
Mean distance to CBD	1.50	miles
Mean distance to nearest high order retail center	1.31	miles
Mean distance to nearest medium order retail center	0.41	miles
Mean distance to nearest low order retail center	0.16	miles
Mean distance to nearest transit line	0.12	miles
Mean time away from home*	164.1	hours
Mean number of trips*	35.7	
Mean distance traveled*	94.9	miles

*During 14-day study period.

$$Y = 66.8 + 19.2x_1 + 0.9x_2 - 57.4x_3 + 0.4x_4 + 20.4x_5 + 25.5x_6 - 132.2x_7 +$$
$$83.9x_8 - 24.1x_9 - 39.1x_{10} + 33.9x_{11} - 34.4x_{12} + 10.7x_{13} + 23.1x_{14}$$

Y is, in this case, the estimate of the number of hours spent away from home. Four of the terms, family size (x_1), educational level of the head of the household (x_3), the presence of one or more workers in the family (x_7), and the presence of one or more school children in the family (x_8), proved to be highly significant (1 per cent level). One other, age of the head of the household (x_2), proved to be statistically significant at the 20 per cent confidence level. The model explained some 60 per cent of the observed household-to-household variation in the amount of time spent away from home. It is interesting to note that only variables in the socio-economic category proved to be statistically significant in this study.

Trip Frequency Study

The total number of trips undertaken by an individual or a household during a given period provides a rough measure of the demand for travel during that period. Possible effects of location factors upon this measure are also of some interest in terms of studies of residential site valuation (see Chapter 8). The total number of trips, excluding movements for the purpose of visiting friends and pleasure driving, was tabulated for each household in the travel sample. The average household made just under 36 trips during the study period.

Application of the general model produced the following estimating equation:

$$Y = 19.9 + 1.7x_1 + 0.2x_2 - 1.7x_3 - 1.9x_4 + 1.7x_5 + 7.7x_6 - 23.9x_7 + 23.2x_8 +$$
$$5.6x_9 - 3.0x_{10} - 3.7x_{11} - 1.3x_{12} - 2.4x_{13} + 23.1x_{14}$$

TABLE 9-IV
ZERO ORDER CORRELATION MATRIX

	1	2	3	4	5	6	7	8	9	10
1	1.00	-.26	.03	.05	.02	.00	.00	.46	.52	.31
2		1.00	-.05	-.18	.00	-.11	-.08	-.11	-.10	-.15
3			1.00	.45	.91	.56	.56	.06	.13	.40
4				1.00	.31	.53	.58	.00	.00	.27
5					1.00	.47	.48	-.01	.02	.35
6						1.00	.53	.10	.12	.44
7							1.00	.16	.08	.29
8								1.00	.81	.66
9									1.00	.64
10										1.00

Variable

1 = Size of family.
2 = Age of head of household in years.
3 = Distance from CBD in miles.
4 = Distance from nearest bus line in miles.
5 = Distance from high-order retail center in miles.
6 = Distance from medium-order retail center in miles.
7 = Distance from low-order retail center in miles.
8 = Total trips/time period.
9 = Total duration/time period.
10 = Total distance/time period.

Note: Values of 0.20 and higher are statistically significant at the 5 per cent level.

Two terms, presence of one or more workers in the household (x_7), and presence of one or more school children in the household (x_8), proved to be statistically significant at the 1 per cent confidence level. One of the location variables, road distance to the nearest low order retail center (x_{14}), was significant at the 20 per cent level. The model explained 49 per cent of the observed variation in the number of trips. The two variables which showed a high degree of statistical significance were of a dichotomous nature, and fell again into the socioeconomic category.

The Total Distance Study

Total distance traveled by an individual or a household provides perhaps the most sensitive measure of transport inputs. If trip frequency may be held to be invariant with distance, then this measure, total miles per time period, should be quite sensitive to differences in residential location. The information on total miles traveled, again excluding travel to visit friends and pleasure driving, was tabulated for the households in the Cedar Rapids travel sample and the following estimating equation was obtained:

$$Y = 189.3 + 5.1x_1 - 1.0x_2 - 49.8x_3 - 96.6x_4 - 39.8x_5 - 79.8x_6 - 26.8x_7 +$$
$$46.5x_8 + 24.2x_9 + 240.1x_{10} - 74.8x_{11} + 99.8x_{12} - 6.0x_{13} - 223.9x_{14}$$

The application of the general model to the data explained only 14 per cent of the observed household-to-household variation in total miles traveled during the study period. Three variables, distance to the nearest transit line (x_{10}), distance to the nearest high order retail center (x_{12}), and distance to the nearest low order center (x_{14}), were statistically significant at the 20 per cent confidence level.

Although locational variables were the only ones to show any degree of significance, the low degree of confidence which must be attached to them, together with the over-all poor performance of the model in this case, leaves much to be desired.

A nonlinear model, of the general form discussed in Chapter 8, was fitted to the data in the hopes of obtaining a better explanation of the observed variation. The estimating equation in this case was:

$$Y = \frac{15.4 x_1{}^{.533} x_2{}^{.093} x_4{}^{.112}}{x_3{}^{.076}}$$

Road distance from the central business district (x_1) proved to be statistically significant at the 1 per cent confidence level, and road distance from the place of work was significant at the 25 per cent level. The model explained 23 per cent of the observed household-to-household variation--a significant improvement over the performance of the larger linear model.

Some Implications of the Present Study

Examination of present theories of urban growth and structure indicates that they are much too general to permit detailed discussion of the effects of transport innovation upon the structure of land values and land use within the city. Chapter 7 pointed out the need for an examination of the problem on a more disaggregative basis in order to provide adequate explanations. The attempts to study empirically the value of access to residential property, reported in Chapter 8, showed the need for more sophisticated structures relating location to the value of residential sites. The failure of simple linear and weighted linear measurements of spatial relationships was also evident.

The relationship of residential site value to the quantity of transport inputs consumed at the site is suggested by current location theory. However, the operational question of what constitutes an adequate measure of transport inputs remains to be answered. If frequency of movement is invariant with distance, the distance traveled per time period should be an adequate measure of transport inputs. If it is not, then more complex multiequation models must be adopted. The empirical evidence bearing on this point is confusing. Some studies report increasing frequencies with increasing distances (perhaps due to incomplete counting of trips); others report decreasing frequencies with increasing distance. Garrison, studying rural areas in Washington State, reported frequencies to be invariant with distance for specific trip types. The empirical studies utilizing the Cedar Rapids travel data (reported in this chapter) tend to confirm his findings.

If indeed, there is no relationship between distance and trip frequency in the short run, some interesting implications arise. First, this would seem to imply that highway improvements cause the generation of no new traffic but only the reassignment of existing traffic within the network. This has an important bearing upon the nature and extent of the impact of highway improvements upon arterial and highway-oriented business types (see Chapter 4). Secondly, the aggregate diminution of trip frequency with distance observed in so many urban traffic studies represents no more than an expression of the increasing number of opportunities for the satisfaction of travel purposes with increasing distance.

If it may be assumed that distance traveled per time period by the residents of a site represents an adequate measure of transport inputs to the site, then this measure should be quite sensitive to locational differences in residential sites. While residential location factors seem empirically to be more important than others, the definitions of location and the simple linear structures utilized in the present studies did not satisfactorily explain the observed variations in transport inputs. The introduction of simple nonlinear weightings to the distance measures improved the situation, but the final performance of the model was still inadequate.

While not brought out explicitly, certain questions which may prove to be of future interest have suggested themselves during the course of the research. For instance, current theory operates on the postulate that differences in location are recognized by the consumer and are important in his decision making. But what if there exists some threshold of indifference on the part of the consumer? Perhaps a hundred yard or even half-mile change in the location of a department store may not be of importance to his evaluation of the situation. Current theory implicitly recognizes an upper threshold but gives no attention to the possible existence of a lower threshold. If it does exist, it might possibly explain some of the puzzling occurrences noted in empirical studies. Also, the topic of weighting of distances deserves more attention than it is currently receiving. Studies of the way in which consumers perceive distance relationships could make important contributions to our level of understanding.

It is felt that the present studies, although preliminary in nature, indicate the existence of a much more complex spatial structure than was previously thought to exist. The existence of this complex structure would seem to dictate a serious re-examination and possible extension of relevant theory. Development of theoretical structures to provide explicit statements on topics of interest would be helpful, and the present lack of operational definitions of many theoretical concepts (e. g. , location) is a great hindrance to empirical testing of these concepts.

At the current level of understanding, it is quite difficult to measure the effect of highway improvements upon the structure of residential land use within the city. The chief difficulty is in making precise statements on how changes in highways cause other changes. The paragraphs above have emphasized the complex set of relationships which contain the problem, and the gaps in our knowledge of these relationships. Only through further intensive research will these gaps be closed and our understanding extended to more adequate levels.

SECTION IV

Analysis of Customer Movement and Retail Business Location

Data on the movement of people are useful in evaluating notions concerning spatial associations of retail businesses. Certain assumptions in current explanations of urban organization set rather restricting limits on possibility of movement within their models. The purpose of Section IV is to evaluate the suitability of these assumptions by directing attention to actual movement characteristics of customers to retail businesses.

The same primary data used in Section III on movement in Cedar Rapids, Iowa, are employed in this section. In this study, however, the focus of attention is on the business site as well as the home base of people moving.

A partial review of studies of location and transportation is presented in Chapter 10, "The Study of Movement Related to the Study of Location." The assumptions in theories and the procedures in empirical studies which implicitly define how people move to obtain the services of retail businesses are identified. Two types of business connections are considered. The first is the spatial relationship of businesses to their customers' home locations. The second is the association of a business with other businesses as established by customers' habits of shopping for more than one item on a single trip.

Chapter 11, "A Description of Customer Trips to Retail Businesses," presents empirical information on movement to retail stores in Cedar Rapids. The spatial arrangement of retail business in the city is established in detail. Characteristics of trips such as length, frequency, and size of center visited are described for each of forty business categories. Analyses of purposes combined in single trips are also presented.

Even in a city as small as Cedar Rapids, movement associations are exceedingly complex. In order successfully to anticipate changes in retail businesses related to highway change, explanations must be developed which explicitly account for multiple-purpose trips. An important shift in emphasis which is suggested is that motives underlying travel habits are based on maximum satisfaction per trip rather than on minimization of travel effort.

10

The Study of Movement Related to the Study of Location

THIS CHAPTER is a review of ideas developed to explain urban organization. First, the essential role of highways in the functional organization of urban communities is identified, with certain elements of this relationship given special attention. Next, summaries of both general land use theory and of related location theory are reviewed from the point of view of movement of people implied by them. Finally, empirical studies of movement and of business location are reviewed with the intention of identifying the empirical procedures necessary for evaluation of the theoretical conclusions developed previously. The reader will note some duplication between these summaries and statements elsewhere in this volume. Although there is duplication, emphasis in this portion of the volume differs from that elsewhere.

Functional Impact of Highway Change

In the past federal aid to highway systems has been restricted primarily to rural highways as connections between cities and to resource roads. Viewed from the activities in a single city, the federal program has been influential in extending the range of external connections. Highway improvements changed accessibility to the city and allowed adjustments in activities in which external traffic requirements were of major importance. However, a large proportion of the funds in the present highway program is to be used to improve roads within urban areas. The impact on urban activities of these new additions to capacity will be great, so important changes in competitive strengths of many urban activities can be expected. The fact that distances traveled within urban areas are much shorter than the external links does not imply they are less important. On the contrary, the high concentration of urban activities is symptomatic of the importance of transfer costs in their operation, and the willingness to pay high rents for particularly accessible sites may be viewed as a substitution of site costs for transfer costs and as a part of the price firms are willing to pay to gain the greatest possible accessibility to related activities.

When a high capacity road of the type being built by federal funds is introduced into this urban system, some establishments stand to gain relative to others. Which establishments benefit will depend upon their location and their need for traffic connections. Changes in operations and locations will depend upon the changes in competitive strength of various types of firms. In order to anticipate changes concom-

181

itant with highway improvement and to group activities by the degree to which they are sensitive to highway conditions, it is necessary to develop information about the structure of urban activities and the role of highway networks in maintaining this structure.

Study of Retail Business Traffic Generation

One approach which is likely to be efficient in yielding information about the role of highways in urban organization is the study of the traffic requirements of urban retail businesses. There are two reasons for this assertion. The first is a pragmatic one: there has always been an interest in and need for studies of business organizations, and as a result there is a large body of knowledge, both empirical and theoretical, related to economics of retail businesses.

The second reason is the belief that the structure of urban movement is so complex that detailed analysis of certain parts is a needed complement to studies of the total movement system; retail businesses and the travel they induce are keys to the entire system. Urban retail businesses are able to pay the highest rents for sites which possess accessibility to the largest segments of the urban region. They take possession of these sites, and other land uses are arranged around them according to their ability to bid competitively for the land. Thus retail businesses form a skeleton of urban organization, and, if understanding of how this type of land use adjusts to road changes is gained, then possible shifts in the entire structure can also be suggested.

Functional Nature of Interconnections

The relationship between the urban road system and the urban land use pattern is a functional one. Many studies tend to view each only morphologically. They are concerned with the percentage of land devoted to roads, to residences, and to other classes of land use. They describe the radical or circumferential pattern of the road net or the neighborhood patterns similar land uses create. An essential point is either missed or given only implicit recognition in such approaches. The essence of urban life is specialization of activities. These specialized activities have space requirements and by necessity must be spatially separated, which creates the differential land uses observed in cities. The activities also require interconnections which may be established by actual physical movement of people and goods or by communication links such as telephone or radio. All the physical movements require facilities, the capacities of which, in part, determine the levels of interconnection possible. One such facility is the road network. Demand for the services of a retail store at any given location depends upon the customer travel utilizing the road network. This movement of the customer from the site of his home to the site of the store is one of the essential interconnections in retail businesses.

The functional approach focuses attention on specialized activities, one facet of which is an identifiable site for the activities and another is the requirement for a set of interactions which is satisfied in part by actual physical movement of persons and things. As the emphasis is on specialization of activities researchers are required to make a very disaggregated and detailed recognition of these activities. Thus theory speaks of the "range of a good, " that is, a range of distances people will travel to obtain a particular commodity. If commodities are aggregated into commonly recognized bundles such as form a drug store or clothing store, it is necessary to identify individual store locations or purposes of single trips to be able to identify the causal relationships between location of a function and use of the road network.

Accepting then that functional explanations of urban structure require detailed empirical data for verification of many of their propositions, one finds such data are usually expensive to collect but occasionally, for one reason or another, detailed data on movement are available. Such is the case in this study. Critical assumptions and procedures of current urban studies which normally escape direct observation because of the expense of data collection may be evaluated with this information. The data available include, for example, direct observations of purposes of trips, combinations of purposes in single trips, characteristic length, frequency and type of center visited in shopping trips for various categories of retail goods, and other data on the actual movement of people in an urban area. The purpose of the study is to suggest which assumptions seem unrealistic in the light of these data and what assumptions might possibly be substituted for them.

Literature Available

Several groups of studies contribute to an understanding of the general spatial organization of urban areas. Each group is addressed to different aspects of urban organization, and different points of view and procedures are used in these studies because of the varied backgrounds of the researchers. Land economists are concerned with urban real estate values and particularly emphasize mechanisms of land rent. Economists concerned with location theory also focus on land rent as a significant variable in urban structure but primarily from a location point of view. A few studies with theoretical statements also are available from researchers interested in transportation. Business research organizations concerned with the pattern of retail trade in cities, especially the central business district versus the suburban shopping centers, contribute mostly empirical studies. They recognize a connection between retailing and parking space with implied or stated relation to traffic volume. Traffic engineers and planning commissions, primarily concerned with providing adequate roads, contribute large quantities of information on movement but generally at a highly aggregated level. Recently, they have become more and more concerned with land uses and how they change with highway improvement.

Certain facets of urban organization appear again and again in a review of these diverse approaches. It is apparent that these recurring elements are the influential variables in both highway utilization and land use systems. Their development is reviewed in this chapter. These elements and the ways of thinking about them have appeared in other parts of this volume and in other summaries of studies with similar points of view. For theoretical statements see Isard (1956), Ratcliff (1949), Troxel (1955), Mitchell and Rapkin (1954), and Berry and Garrison (1958b). Summaries of empirical studies include: Weiss (1957), Schmidt and Campbell (1956), Barkley (1951), Voorhees *et al.* (1955), and Kelley (1956).

The theories of business location which are summarized in Section II of this volume are reviewed again in this section. This summary is modified by a point of view adapted from Mitchell and Rapkin (1954). Their studies of Philadelphia were explicitly designed to integrate information on urban transportation with information on urban land uses. Troxel's (1955) comments on a theory of the movement of people are a third and stimulating source of ideas expressed here. There is considerable merit in bringing these diverse approaches together. Each contributes information and ideas to the other in areas where the individual studies are vague or where little emphasis has been given.

Location theory applied to urban retailing appears to bring the phenomena into some general order and further demonstrates that the structuring of this type of activity is one example of the operation of spatial factors probably affecting all social activities to some degree. Such a general approach is required if understanding is to lead to some measure of control. At the same time, emphasizing elements of movement within the system introduces a new dimension to these theories, a dimension which can be used to measure their effectiveness. The elements of movement identified by Mitchell and Rapkin are used to evaluate some of the concepts of location theory which have to do with defining how people move within urban areas. Troxel's comments on movement suggest the kinds of adjustments which are needed in general theories of location and in the applied procedures of researchers.

Definitions

To help keep the discussion clear, the following paragraphs explain the meanings attached to some common words. Our emphasis is on retail industries. An *industry* is a group of firms all of which produce nearly the same bundle of goods or perform approximately the same function. A *firm* is an organization of specialized activities designed to perform that function; *firm* is generally thought of as a business term. The *household* is the analogous organization for establishing a residence or home site. The firm or household does not define location; rather, they create *establishments* which define the business locations or residences. A firm may consist of several establishments perhaps with different functions but nevertheless organized into a single system. For example, a grocery chain may have the offices of the management in the central business district, warehouses near external transportation facilities, and retail outlets distributed according to the distribution of population. On the other hand, an establishment and a firm may be one and the same and may also represent a whole industry in the city.

Each establishment is the site of a bundle of activities. These specialized activities have certain requirements for interconnection with other establishments. The interactions or interconnections consist of *communications* and *linkages*. *Communications* are information contacts by letter, telephone, and the like. *Linkages* are movements of people and goods between establishments. We are concerned in this study primarily with linkages consisting of customers who, by their travel patterns, connect retail establishments with one another and with residences.

The sum total of residences, business locations, and other public or manufacturing establishments plus the transportation *facilities*, which are the channels or routes of the linkages, establish the *land use pattern* of a city.

Review of General Theory

There is an essential order to the apparently haphazard arrangement of land use. One may observe the concentration of major urban activities such as retailing, manufacturing, and so on, into functional areas. The same general tendency appears in all cities despite minor differences due to topography, different social attitudes, and the like. The mechanism of urban society is by and large economic and R. M. Haig, a land economist, is generally credited with the first extensive theory of urban organization (1927). His work, although amplified by such students as Ratcliff, Wendt, and others, is still fundamental in present day studies. Basically, such studies hold that

Utilization of land is ultimately determined by the relative efficiencies of various uses in various locations. Efficiency in use is measured by rent-paying ability, the ability of a use to extract economic utility from a site. The process of adjustment in city structure to a most efficient land use pattern is through the competition of uses for the various locations. The use that can extract the greatest return from a given site will be the successful bidder. The outgrowth of this market process of competitive bidding for sites among the potential users of land is an orderly pattern of land use spatially organized to perform most efficiently the economic functions that characterize urban life [Ratcliff, 1949].

Rent
Economic utility is gained when a site is used as one of the factors of production of some enterprise. Land value is the capitalization of the utility of a site. It will reflect the greatest productivity the most appropriate activity can achieve or rather the productivity which the landowners and entrepreneurs estimate will be achieved during the life of the establishment.

The role rent plays in an enterprise has been a much debated question in economics, but the intricacies of these arguments will not be presented in this brief summary. The view accepted here is that rent is a cost factor which helps to determine price and not that rent is a residual determined by price. A site has no value except as it is developed by some activities although, of course, the value may be for some anticipated activities. The utility that is added by a site, which cannot be explained by the capital and labor applied to it, is primarily locational. This is especially true for urban sites as compared to agricultural sites which may have both a locational value and a potential physical fertility. Rent is a charge landowners can levy as a substitution for savings in transportation costs.

Rent appears as the charge which the owner of a relatively accessible site can impose because of the savings in transportation costs which the use of his site makes possible. The activities which can "stand" high rents are those where large savings in transport cost may be realized by locating on central sites where accessibility is great. . . . While transportation overcomes friction, site rental plus transportation costs represent the social cost of what friction remains [Haig, 1926].

The general social aim is land use organization which will minimize this total cost. This is accomplished by each establishment minimizing its transportation costs and then being ranked by rent-paying ability. In allowing the firm with the greatest rent-paying ability to bid for and gain the most accessible sites, those establishments with the largest combined rent and transportation costs gain locations which keep at a minimum the total social cost of travel. In this interpretation the social costs of distance are at a minimum, and this result is not in conflict with the individual firm's desire for maximum revenue.

Minimum Transportation Costs

As rent-paying ability is the determining factor in the location of an establishment, the relationship between transportation costs and rent is the element of the land use theories which is most pertinent for evaluating the impact of highway change.

Linkages, to which some kind of transfer costs are attached, are necessary to a firm and vary according to the bundle of activities carried on by the firm. The cost of transportation will vary accordingly. In retail stores the most important linkages are with customers' home sites. The customers will be a given set of people chosen from all people in the urban area. They may be grouped in a single sector of the city, or they may be diffused throughout the entire area or may come from beyond the city limits. From the firm's point of view, the objective is to maximize net revenues, and this is accomplished only when transportation costs are minimized and a bid for the best possible site is successful. In the case of retail stores, the cost of customer travel is the most important transportation cost, so locations are sought to minimize this travel.

Intensity of Land Use

Of course other factors of production affect rent-paying ability. As rent is thought of as a charge per unit area, the intensity of land use by various activities affects the rent-paying ability of the firm. An establishment, the operation of which requires large internal space, will ordinarily have less rent-paying ability than a space-intensive establishment. Furniture stores, though their operators would be pleased to locate in the most accessible site, are generally on the outskirts of the intensively built central business district. A women's clothing store or a jewelry store on the other hand has higher intensity margins and can enter this area. Doctors, dentists, and other professionals can generally use space intensively, and this condition helps them to compete for central business district sites, albeit normally in upper floors of downtown buildings. The location in upper floors is probably because such firms do not require display space.

Other internal firm characteristics affect rent-paying ability. The rent-paying ability of a jewelry store or fashionable dress shop may be greater because of the markup possible on their goods than, for example, a grocery store's. Markup is small and competition strong among grocery stores where there is very little to distinguish between the commodities offered while a fashionable shop may gain a value in exclusiveness part of which may be used to obtain a certain location or association with other establishments.

The schedule of substitution between production factors allows flexibility in location and is one reason why similar type stores are found in many different rent level areas. Consider the jewelry store where ordinarily the large markup reflects an adjustment for slow turnover of goods. By moving to a very accessible site, the firm increases its rate of turnover and is able to pay a higher rent while still making the same profit, but such a substitution most likely would require changes in other characteristics such as credit or advertising policy as well. A perfect substitution obtains when jewelry stores are perfectly competitive. Any attempt to hold a choice location and gain excess profits because of the ability to operate more intensively would be lost to the landowner as rent. In actuality, the imperfections in competition which exist allow variation in choice of location for firms of similar types.

Scale of Operation

Another internal firm characteristic influential in firm locations is the minimum scale of operation. Some activities have such a large minimum scale of operation that only one or at least only a few firms of that kind can locate in a metropolitan region. Department stores needing a great volume of trade, on the one hand, and highly specialized stores such as one selling only maps, on the other hand, need a very large general population to support them. In contrast, the corner drug store needs only a few people in order to recruit enough customers for its support.

Consider the different distribution of customers for the small, downtown specialty shop and the small, neighborhood store such as a corner grocery. Even postulating the same number of customers to each type of store, for the first, the customers' home sites are scattered throughout a large area, while for the second, they are concentrated into a particular sector of the city near the store. The total cost of obtaining the services of the specialty store involves greater transportation costs than do trips to the grocery store. Improvements in highways which change general access throughout the urban area would have a greater direct effect on the specialty store. If the small grocery was located so that access to the immediate neighborhood was changed, a direct effect would be registered; but more than likely an indirect effect, caused by a change in competitive strength, say between the small store and supermarkets, would occur for the grocery.

Monopolistic Elements in Location

The ability to pay rent will depend upon competition with firms of similar lines. The ease of entry may be important in determining whether existing firms can maintain their present locations in competition with other land uses. This point is not emphasized in most theories which tend to consider conditions of perfect competition.

It is significant that Chamberlin (1933) considered locational factors in his theory of monopolistic competition. Monopolistic elements in firm operation may be obtained because of locational elements. For example, the effort required to go to the next nearest competitor may be capitalized upon. This line of reasoning is not considered with the marginal production approach to rent. The monopolistic element in location involves similar firms (within the same industry) and not other types of activity. Thus a firm may raise its price to the degree people are willing to pay, rather than go farther, and the firm may not lose its volume of trade, but the excess so gained will still be lost as rent.

The important distinction between competition by several firms within an industry and competition between industries is overlooked in many interpretations of location studies because restricting assumptions made early, and many times only implicitly, in the analyses as to what sector is to be considered are overlooked. Invalid inferences are sometimes made if the distinction is not recognized.

It is a mistake to believe that because land is limited and immobile monopolistic policies can be established in urban areas. The firm locations themselves are not immobile in the long run, and the firm best able to bid for a site will be the one to obtain its use. This factor establishes the most efficient use among all choices for each parcel of land. Competition within a given industry tends to force the firms to employ the most efficient methods and scale of operation. These methods and scales of operation need not be the same for all members of an industry; in fact, they would not be since establishments in areas of different rent levels have different factor cost ratios. For most industries, location near the center of a city means greater intensity and, commonly, a large scale of operation as well.

Rate of Adjustment in Location

The most efficient land use pattern will result from competitive choice of location. Actually, obvious inefficiencies in urban organization exist. The urban system is dynamic, but certain facets, particularly land investments, are slow to change following growth of population or technical innovations. The planning period for a large building, for example, is long and involves great investment. Conditions may change radically, but the least cost procedure for the owners of the building may be to keep the present accommodations in operation at low efficiency rather than tear the building down and start again. Such decisions depend on the circumstances. As the city is always in a dynamic state of change and the planning period is long for many activities and facilities, inefficiencies will always be apparent.

Summary

To reiterate, the land use system is an ordered economic phenomenon which is the result of competition between firms capable of paying different amounts of rent. The rent-paying ability of a firm depends upon the internal firm characteristics, the schedule of substitution between factor costs, and the minimization of transportation costs. Location is relative to other establishments only where linkages are required. A good location can become a poor one because of changes in internal activities which require different linkages or because of shifts in the location of establishments with which the firm interacts.

Location Theory

Location theory has dealt with particular problems of location of firms and industries. Principal emphasis has been on manufacturing and agriculture. Von Thünen (1826) developed the first principles, the importance of which were underscored and expanded by Lösch (1944) and recently again by Isard (1956), to mention a few among an increasingly large group of modern researchers. These developments are traced in detail in Section II of this volume. Location theory essentially accepts the ideas of land use theory but goes beyond identification of the forces involved and attempts to explain the shape of the land use pattern which emerges. A major contribution of these writers is their convincing argument that the location of economic activity is the result of an orderly process subject to measurements.

The studies have approached the problem at two levels, microeconomic problems of the location of firms within industries and macroeconomic location of industries within the entire economy of a region. We have described the variables which are important in location of urban activities as developed by real estate economists, particularly the role played by the rent-paying ability of firms and industries. Essentially the same variables have been developed independently by economists and geographers interested in agricultural and manufacturing location problems and in systems of city locations but with transportation receiving a more explicit role.

Competition Within Industries

Studies, such as those Hoover (1948) describes and expands, consider competition between firms selling the same products. The firms are considered without reference to other types of businesses. Point locations of the competing firms are given, and the problem is to determine how the market is to be divided. The cost necessary to reach any point with the product is the deciding factor. Hoover recognizes two general kinds of costs, production costs and transportation costs. Factor substitution between the two types is important in determining the share of the market obtained.

Size of Market Areas Determined

The analysis is extended in two ways by other studies. One is to consider the number of firms as variables and determine the optimum number of firms which will be needed to supply the market best. This will depend upon the optimum size of firms, again considering the free substitution of transportation costs and production costs. In perfect competition and because of equal availability of all factors, the size of the firms will be the same and all excess profits disappear (see Lösch, 1944).

General Equilibrium System

Another dimension possible is to include all other businesses in the analysis. To obtain the optimum location, a firm will now find it necessary to compete with other land uses for space. Rent appears as a factor price, and the substitution schedules now possible include production costs, transportation costs, and rent costs. The number and distribution of the firms in each industry are unknowns sought. A general equilibrium system is necessary to determine the location of any one firm. Lösch and Isard stress the need for such a general system, and each attempts to formulate one. The systems they suggest include the elements of urban location as a special case. Although the systems resemble the real estate economists' explanation or urban land use, they arise from different sources.

Von Thünen

Certain prominent notions in modern location theory are from the works of two Germans, von Thünen and Christaller. Von Thünen (1826) contributed an analysis of the important role the rent-paying ability of an activity has in determining its location. References to von Thünen are absent in real estate literature perhaps because he dealt with agricultural location. He analyzed the problem of what products a farm should produce given its location with respect to a single market. He showed that as products had different transportation costs attached to them they would produce different net returns at a given distance from the market. The choice of location would depend upon the rent-paying ability of each product, the crop with the maximum rent-paying ability for a given location being the one grown. Because other things were held equal, the rent-paying ability of a crop was directly substitutable for the transportation costs. Von Thünen showed that, given a homogeneous plane, economic specialization will still develop and that there is a locational equilibrium for the entire system.

Christaller

Walter Christaller (1933) was concerned with the spatial arrangement of central places, towns and cities, related to consumers distributed over an area. The relationship which exists between central place and consumer is a functional one-- the supply of goods and services. One important implication of Christaller's works, and later in the study by Lösch, who analyzed the notion in greater detail, is the existence of a hierarchical arrangement of centers. Each class of centers possesses all the functions of the classes below and a set of higher functions as well, the largest center possessing the full range of functions. In order to obtain the use of the highest function, consumers must travel to the highest center or have goods distributed from these points. The shortest distances traveled are to the lowest order centers where only the lowest order goods are obtained.

Hierarchical Classes of Shopping Centers

The similarity between a network of cities (from the great metropolis down to the smallest hamlet) and a network of shopping centers (from the central business district down to isolated stores) is apparent. Berry (1958) takes advantage of this parallel to apply this understanding of city systems to business center systems within cities. He points out that the hierarchical class system is the important contribution of central place theory, not the hexagonal pattern of market regions obtained by Christaller and Lösch, who assumed equal population distribution and other restricting assumptions. Isard suggests a more probable shape of functional regions around central places in the last chapter of his book (Isard, 1956).

The Role of Transportation

Movement is recognized in these theories of land use and location by reference to cost of travel, transportation or freight rate, distance inputs, and by more oblique references such as the size of the market region and the range of a good. In Isard (1956), general equations for the space-economy boundary condition between any two producers (or distributors) are established by equations containing a term for marginal production costs (excluding transportation) and distance and transportation rate elements. The equations "furnish the boundary (indifference) conditions dividing a market domain between any two producers, each boundary representing a locus of points of equal delivered prices" (Isard, 1956).

Isard recognizes the similarity between his formulation and that of Lösch. "In Lösch's theme, the delivered price is equal to average production costs (factor price) plus transport cost (the consumer is responsible for the transportation of the product he purchases). Hence, the boundary (indifference) line between market area of any two producers (is) a locus of points of equal delivered price" (Isard, 1956). The boundary depends upon a distance input, transportation rate, and average factory price.

Note that in neither scheme do people move beyond the least cost point to obtain a good. Besides the explicit assumptions stating the boundary conditions, two basic assumptions underlying the whole analysis also control the possible range of movement. First, complete information is available to all economic agents in the theories; the consumers make no mistake due to lack of information. Second, there is perfect substitutability between producers. Peculiar attributes of the central places or the entrepreneur do not affect movement decisions. Minimum costs necessary to fulfill demands are the only motive for establishing movement patterns. Combinations of purchases on single trips are not considered. Such behavior by consumers might suggest another motive in movement, that of maximizing net returns of a trip.

Berry and Garrison describe the idea of a range of a good:

This range marks out the zone or tributary area around a central place (urban center) from which persons travel to the center to purchase the good, ". . . a product of the simultaneous spatial effects of all the factors of demand and supply involved in the purchase of central goods and services" (Christaller, 1933, page 20). The *upper limit* of the range is the maximum possible radius of sales. Beyond the upper limit the price of the good is too high for it to be sold, either because of the increase of price with distance until consumers will no longer purchase the good (the *ideal limit* where demand becomes zero), or because of the greater proximity of consumers to an alternate competing center (the *real limit*). The range also has a *lower limit*, that radius which encloses the minimum numbers of consumers necessary to provide a sales volume adequate for the good to be supplied profitably from the central place [Berry and Garrison, 1958c].

Lösch and Isard postulate the smallest possible limit of a good that is necessary to provide a profitable sales volume at the center. The modification allows surplus profits to exist, profits which in reality are divided between greater rent cost and excess profits for the firms. Again we wish to emphasize the actual patterns of movement implied by this idea of a range of a good. Movement is confined to a region bounded by a zone of indifference points and points where demand is zero. Again the discussion is restricted to one good at a time. There have been references in the literature concerning the association of businesses, but they have been restricted to general statements or empirical observations (see Berry, 1958, p. 108, and Hoover, 1948, p. 123). An explicit statement of the effect of combining the transportation costs of various goods is not available.

Transportation Studies

We shift now from a discussion of how transportation considerations have entered theories of land use and space economy in order to review some recent studies of transportation which involve considerations of land uses. Studies of transportation considered here fall into three groups. Mitchell and Rapkin's (1954) study of urban traffic as a function of land use summarizes many new ideas on approaches to the problem and possible measures which can be made to develop explicit information on how people move in urban areas. Several studies have accompanied and followed this work; they form a group because of the new method of data collection and assembly that they employ.

A second group consists of studies of central business districts; Weiss (1957) presents a good summary of these. Studies of suburban shopping centers and parking studies, mainly by business research organizations and planning commissions, are also considered.

Traffic engineers and planning commissions have developed a large amount of empirical information on movement of people in urban areas. These studies form the third group.

The studies of transportation are nearly all empirical, many done under the pressure of finding a solution to a traffic or parking problem. We review them more for methodology than for information. Our purpose is to demonstrate how empirical procedures contain implicit assumptions on the manner in which people move. These assumptions may or may not agree with our notion of actual movement. Many times they are in conflict with those notions and even with other empirical data. We hope to identify elements of movement patterns which most commonly become obscured because of the methods employed in the empirical studies.

Studies of Central Business District and Related Subjects

Weiss (1957) divides studies of the central business district into four types: (1) retail sales attraction, (2) land values, (3) daytime population, and (4) functional classification.

The usual question in central business district studies concerns trends in the competitive struggle of the central business district with suburban shopping centers. Retail sales attraction studies show trends in percentage of retail sales volumes in the central business district compared with the entire metropolitan area. Little reference is made to movement of people although it is certainly implied. Weiss suggests the reason central business districts are receiving a smaller percentage of trade is that growth has been taking place "at greater average distance from the central business district than heretofore and consequently sales have been more dispersed," that is, people do not travel to the central business district because other dispersed sources of supply are available nearer at hand. Therefore, total movement for shopping is more dispersed.

Land value studies make more direct reference to movement. Hoyt (1933) relates land values in Chicago's central business district with technical changes in transportation. William-Olssen (1940) recognizes relationship between heaviest pedestrian traffic and the highest shop rent. The characteristics, such as length or origin of the trips which brought people to these locations, are not considered. Many studies comment on the heaviest pedestrian traffic as a measure of the center of the central business district. Breese (1949) analyzed the composition of daytime population in downtown Chicago. He used very gross categories of purpose: employees, permanent residents and hotel guests, and pedestrians. The pedestrians of off-peak hours were considered to be shoppers because of their concentration around the great department stores.

To associate land uses by linkages in movement requires a detailed breakdown of both movement and land use, and the trend in functional studies of central business districts is toward more detailed land use classification. Rannells (1956), Ratcliff (1953), and Weiss (1957) all suggest more rational and more detailed land use classifications. Unfortunately, their emphasis is on land uses which display similar space-using characteristics. No thought is given to groupings by similar bundles of linkages attached to each activity, but this may be expected since their emphasis is not on movement. Ultimately, however, the changing land use of central business districts must be linked to changing travel patterns. Classifications that ignore this connection, even though they are detailed, may very likely obscure pertinent variables.

Origin and Destination Studies

Origin and destination studies contribute large quantities of data on movement. Barkley (1951) and Schmidt and Campbell (1956) provide extensive bibliographies and summaries. After a long period of development this type of study is becoming more useful in the analysis of the relationship of movement to land uses, and many origin and destination studies are now combined with detailed land use information. Carroll and Bevis (1957) describe the procedures predicting travel generation in the Detroit area. Hall (1958) describes purposes of trips in detail. This is the key to better understanding of urban traffic, not measures of volume of traffic only but of purposes as well.

Two procedures continue to obscure functional relationships of movement in origin and destination studies. The first is the common classification of purposes into six to ten categories which prove too gross for many purposes. This difficulty might be overcome if a few studies which had detailed information on purposes of trips were to establish a relationship between their results and those of origin and destination studies. The broad coverage by the latter of many cities of various sizes might increase the generality of detailed but restricted studies on movement. The present study falls into this category, but this task cannot be undertaken here.

Second, functional relationships are obscured in origin and destination studies by assigning a trip only one purpose. Many trips, in fact perhaps most trips, have more than one purpose. This failure cannot be overcome in most studies even by going back to original data-collecting schedules.

Customer Buying Habits and Attitudes

A group of studies which contributes much to the description and understanding of travel patterns consists of those concerned with shopping habits, for movement plays a major role in the studies. Voorhees, Sharpe, and Stegmaier (1955) present a summary of major contributions. Other studies include Kelley (1956), Jonassen (1955), and Pendley (1956). Some of the studies employ interviewing techniques developed by sociologists. Emphasis is upon motivation for shopping trips, such as attitudes toward the convenience or inconvenience connected with shopping in the central business district versus suburban shopping centers.

This approach to travel behavior is encouraging because it uncovers motivations which shape travel habits. "A wide selection of goods" and ability to "accomplish several errands at once" rank high as motives for downtown shopping (Jonassen, 1955). Troxel (1955) suggests that travel habits are formed by a desire to maximize net return of travel over a period of time, and customer buying habits studies can supply the information needed to analyze this assertion. The implications of location theory are that increased travel costs may be incurred if expectation of greater returns from trips is possible. Baumol and Ide (1956) develop a similar line of reasoning.

A criticism of these studies is that the classification of purposes of trips into categories is not designed to show differences in travel behavior. Thus, a classification of businesses into retail goods and retail services is based on an operational characteristic of the businesses, although a more suitable classification of businesses would be one based on customer trip characteristics commonly associated with retail establishments.

Components of Movement Structure

The main contribution of Mitchell and Rapkin (1954) is the identification of the characteristics of movement and the classification of movement into systems based upon a variety of criteria. They considered a great number of trip types that can be recognized in urban areas and suggested terms which would facilitate the use of them in research. Establishments, home base, work base, linkage, destination, destination area, and many more common expressions were given explicit definitions.

Trips are identified by terminals which include *bases* and *destination*. Home, work, and school are considered *base* establishments because of the regular, recurring visits persons make to these sites. *Destinations* are places where *transactions* take place. A transaction is thought of in a broad sense. It is any event of significance to an individual which occurs in the course of a trip. There are destination points or establishments which are precisely definable locations. A *destination* is defined as a limited area where a number of stops are made at different destinations. It may be desirable to define this as the smallest area which includes the establishments and links between them.

Trips designated by their terminals include *interbase movement* (e.g., from home to work and back), *round trip movement* (from one base to one or more destinations), and *compound interbase movement* (two or more bases and destinations).

The above terms refer to individual trips, but mass movement is also considered. There are mass movements of assembly and of dispersion. An assembly area is a relatively small region where trips may terminate, the most prominent example being the central business district. As most movement by people in urban areas is two way, assembly areas imply a related dispersion area. It is obvious areas of assembly may consist of overlapping destination areas of all individuals moving during a given time period.

Data Collection and Evaluation

Where do all these definitions lead us? Consider two points. First, the simple systems described above are subject to measurement. Several methods of measurement are suggested. Interviews of a sample of people at home, at work, or in stores, depending upon the needs of the researcher, may be designed to yield information regarding these movement systems. Mitchell and Rapkin present the results of such interviews conducted in downtown Philadelphia retail establishments: a department store, a theater, a retail specialty shop, and a small service establishment. They wished to establish and compare with other retail outlets the kinds of trips and associations characteristic of these four types of businesses. The results showed considerable difference in the linkages of each store.

Another method of obtaining such data is by travel diaries like those used by the Traffic Audit Bureau (1947, 1950). The data gathered by that organization in Cedar Rapids, Iowa, are the basis for the study in Section III which centers on the home base and for the descriptions given in Chapter 11 of this section where the emphasis is on the retail business establishment. Identification and classification of movement thus allow measurement and analysis to follow. The detailed definitions of the elements being studied are justified, in the first instance, as necessary steps in empirical research.

Theoretical Implications Suggested

The second point to consider is how to relate such empirical abstractions to existing theories of movement. In this case an interpretation is possible. The assembly areas can be regarded as shopping centers and their corresponding dispersion areas, the market area or hinterland of each center as described in location and central place theory. These theories applied to hierarchies of urban retail shops imply specific movement characteristics of customers traveling to each class of center. These characteristics are observable within the framework of the empirical procedures outlined above.

The associations of businesses linked by customer travel patterns are also observable. These associations may be particularly strong for certain types of store, for example, one would expect department stores and clothing stores to be so linked. The implications of such associations are important in interpreting the minimum travel-cost criterion imposed on movement in theory. An individual may, in actuality, pass directly by a clothing store in a lower order center and visit an identical store in a higher center where a department store is located. Such movements are denied in theory and several theoretical interpretations of this discrepancy are possible, but first it would help to establish the importance, i.e., the magnitude, of these suspected business links through customer travel habits, by empirical evidence. Customer travel habits are the subject of Chapter 11.

Summary

This chapter has been a review of ideas which have been used to explain urban organization. Emphasis has been on the role of movement in order that some method of evaluating highway change which is based on sound and general theory may be sought. Second, empirical procedures necessary to evaluate these ideas have been identified. The relationship between land uses and highways is a functional one and has very disaggregated features which are likely to be important. That is, research on travel behavior of individuals is a necessary supplement to studies of mass movement if road changes are to be placed in proper context in terms of present day theory of urban organization. Individual movements can be observed only by sampling the mass of movement in urban areas. Which questions are important enough to be approached first and methods of answering them have been suggested in this chapter. Analysis of them is made in the next chapter. Specifically, two of the points developed above are the subjects of Chapter 11. They are to determine whether:

1. The movement between customer home locations and stores in each level of the hierarchy of shopping centers will describe the hinterland or market range for each level, if indeed they exist as theory suggests.

2. Associations of trip purposes may be used to evaluate the importance of linkages between retail establishments and other establishments, both retail or other, in the location of shopping centers and types of retail stores.

11

A Description of Customer Trips to Retail Business

THIS CHAPTER investigates trips to retail stores within an urban area as a means, first, of evaluating the idea of a hierarchy of shopping centers and, second, of ascertaining the degree to which shopping purposes tend to be grouped into single trips. Together, these assist in explaining how highway improvements result in changed patterns of travel and location patterns of land uses.

The point in the preceding paragraph may be made more exactly. The study gives attention to two distributions, each having complex characteristics associated with it. One distribution, that of business establishments, has characteristics such as the grouping of establishments into business centers; this distribution may be described empirically and interpreted by location theory. The other distribution, that of trip purposes, has characteristics such as the association of several purposes into a single trip or the concentration of many trips at certain locations. What are the relationships between these two distributions? This is the simple question the study sets out to answer, but the characteristics of the distributions are so complex that no simple answer will suffice.

The question is significant because the bundle of trips associated with every kind of land use makes up the elements which connect use of the road network with use of sites by the multitude of urban activities. Customer travel to retail stores is one example of such connections. Improvements in travel facilities will change travel patterns. Estimates of how these travel patterns will change and how their change will affect the level and organization of other activities depend in part on the ability to understand the characteristics of these connecting links.

The results of this study do not make incisive statements possible. However, the study establishes the nature of the research problem and presents a method of approaching it. New information is presented which suggests much about the problem and the character of its solution.

The data used are from the Cedar Rapids, Iowa, study described in Chapter 9. However, in the present chapter reference is chiefly to retail business establishments rather than to the residence. Routes traveled, the frequency of trips for various purposes, the combination of purposes on trips, and other descriptions pertinent to the problem are developed from the Cedar Rapids data.

The chapter is divided into three sections: the first is a general discussion of the Cedar Rapids study area, the second is a detailed description of the retail business structure of the city, and the third section presents a description of customer

Figure 11-1. Cedar Rapids looking northeast over the central business district and the northeast sector of the city. Photograph courtesy of the Cedar Rapids Chamber of Commerce.

trip characteristics and certain general suggestions and conclusions from the analysis.

Study Area

The comments below serve to orient the reader to the study area. It is stressed that Cedar Rapids seems to be a fairly typical example of urban communities in this country. In particular, the notion that land use and traffic are typical assists in arguing generality for the results of this analysis as well as the results of the analysis in Chapter 9.

Cedar Rapids

Cedar Rapids is located in Linn County, Iowa, in the southeastern portion of the state. The population within the corporate city limits was 72,296 in the 1950 Census. Except for Iowa City, 26 miles to the south and with a population of about 27,000, Cedar Rapids is the only large town for a distance of 70 to 100 miles in any direction.

Cedar Rapids serves as the central market and supply point for a productive agricultural region. However, manufacturing is also an important part of the city's economy, as a recent count shows over 200 manufacturing firms employing approximately 21,000 persons and five firms employing over 1,000 employees each. Manufacturing is diversified; food processing, manufacturing of electronic equipment, and manufacturing of heavy farm and construction machines are especially important in the complex of industries.

Notes on Urban Transportation

Cedar Rapids is served by five railroads. U.S. Highway 30, a main east-west route, passed directly through the town at the time the data on movement were collected in 1949, but today the highway partially by-passes the town to the south.

The proportion of traffic approaching a city which is destined for the city itself, or conversely, the proportion which would like to by-pass the city, is a function of city size. Cedar Rapids is typical in this respect to cities in its population group as shown by Table 11-I.

TABLE 11-I
PROPORTION OF TRAFFIC BOUND TO AND BEYOND
CITIES OF VARIOUS POPULATIONS

Population Group	Number of Cities	Traffic Bound to the City (Per Cent)	Traffic Bound Beyond the City (Per Cent)
Less than 2,500	6	49.3	50.7
2,500 to 10,000	6	56.7	43.3
10,000 to 25,000	3	78.1	21.9
25,000 to 50,000	5	79.0	21.0
50,000 to 100,000	2	83.8	16.2
Cedar Rapids, Iowa* (population 85,000)	–	85.1	14.9
100,000 to 300,000	2	81.6	18.4
300,000 to 500,000	2	92.8	7.2
500,000 to 1,000,000	1	95.8	4.2

Source: Schmidt and Campbell (1956), p. 79.

*Source: *Cedar Rapids Origin and Destination Study* (1958). Population estimate is for Cedar Rapids in 1957.

The availability of parking spaces has been shown by many studies to be an important consideration in customer's choice of shopping, e.g., Jonassen (1955) and Voorhees *et al*. (1955). In this respect the stores in the central business district of Cedar Rapids may have a better competitive position than stores in comparable position in average cities of this size (Table 11-II).

TABLE 11-II

AVAILABILITY OF PARKING IN SELECTED CITIES AND BY SIZE OF CITY

City	Population of Urban Area	Parking Space in Business District	Parking Spaces per 1,000 Persons
Baton Rouge, La.	90,000	3,300	37
Cedar Rapids, Iowa	80,000(estimated)	4,153	52
Decatur, Ill.	77,000	3,684	48
Hamilton, Ohio	51,000	1,526	30
Quincy, Ill.	42,500	2,307	54
Hutchinson, Kans.	34,500	4,454	129

Population Groups	Number of Cities	Parking Space per 1,000 Persons
500,000 and over	2	12
250,000 to 500,000	6	28
100,000 to 250,000	8	42
50,000 to 100,000	2	57
25,000 to 50,000	3	66
Less than 25,000	5	90

Source: Bartholomew and Associates (1950a), p. 23.

Table 11-III presents a comparison of travel purposes in Cedar Rapids with travel purposes in cities of various sizes where for the most part the categories of purposes were comparable. The breakdown of trips to retail establishments for

TABLE 11-III

TRAVEL WITHIN URBAN AREAS: TRIP PURPOSES BY DESTINATION

City	Population of Corporate City, 1950	Time of Study	Work	Busi-ness	Medical-Dental	School	Social Recreation	Eat	Shop-ping	Serve Pas-senger	Change Mode	Home
San Francisco, Calif.	775,357	1955	27.3	4.1	1.1	2.8	12.2	1.9	6.2	5.5	1.2	37.7
Houston, Tex.	596,163	1953	21.0	4.0	1.0	4.0	11.0	3.0	9.5	6.0	0.9	40.0
Dallas, Tex.	434,462	1951	18.9	4.3	1.0	4.9	10.8	1.7	10.1	7.3	0.6	40.4
Portland, Ore.	373,628	1949	20.9	7.4	1.0	0.7	15.4	2.1	9.3	5.0		38.2
San Antonio, Tex.	408,442	1956	17.4	5.8	1.7	5.8	9.1	2.2	8.8	8.5	0.5	40.8
San Diego, Calif.	334,387	1953	14.8	14.0	0.8	3.8	9.3	1.7	8.0	7.0	1.8	33.4
Tulsa, Okla.	182,740	1955	18.4	5.6	0.8	3.2	15.2	2.1	10.1	6.9	0.3	40.2
Sacramento, Calif.	137,572	1948	33.6	4.8	0.7	2.6	10.1	1.7	8.1	6.7	0.2	31.4
Tacoma, Wash.	143,673	1949	21.8	6.0	1.2	2.2	13.7	1.7	7.7	4.9	0.7	40.0
Erie, Pa.	130,803	1950	26.7	1.4	0.9	0.4	15.5	1.0	9.1	5.0	0.2	39.8
Fresno, Calif.	91,669	1955	15.0	11.3	0.8	3.9	10.3	2.5	9.5	7.0	0.3	33.6
Cedar Rapids, Iowa	72,296	1958	22.4	10.6	0.7	1.1	6.2	1.5	7.6	11.5		38.4
Stockton, Calif.	78,853	1953	15.7	21.2	0.4	1.1	6.4	2.8	8.4	12.5		27.4

Source: *Metropolitan Area Traffic Studies* of the cities listed.

purposes of shopping, eating, medical-dental visits, and so on shown in Table 11-III is one classification of trips which is available for comparison from city to city. Percentages of trips devoted to these purposes in Cedar Rapids are approximately the same as found in other cities.

Traffic flow in Cedar Rapids is similar to that in other urban communities. A transit bus system is in operation but automobiles comprise most of the intraurban traffic generated. Figure 11-2 is a traffic flow map of Cedar Rapids which clearly defines the major arterials. This may be compared with Figure 11-3 which shows in detail the location of all retail shopping centers down to single isolated stores. The orientation of shopping facilities to traffic flows and vice versa is quite clear.

The bridges across the Red Cedar River restrict the cross-town traffic flow to five routes, four of which pass near or through the central business district. One might be led to think this necessary concentration of traffic into the central area may be greater than that of most cities. However, a number of cities have similar barriers running through them. Commonly, the barrier is a river such as in Cedar Rapids, but railroad lines or limited access highways may also be barriers. Also, much of the traffic within urban areas is bound for the central business district and would enter the area regardless of route choice.

First Avenue, the main north-south route, has a 120-foot right-of-way which is an asset to handling traffic in the central area. The main commercial establishments, however, are not on this wide street. William-Olsson (1940) commented on a similar relationship between the main arterial and most concentrated shopping areas in Stockholm, Sweden, where the businesses have located near, but not on, the through traffic channels.

North of the central business district along this street there is a development of retail stores which is oriented to the road but which displays a different character than the central business group. Here is the most developed case in Cedar Rapids of the familiar string street business grouping found in most cities.

Notes on Land Use

Figures 8-1 and 8-2 in Chapter 8 show major land use patterns and distribution of population within the city. Table 11-IV summarizes the land uses in tabular form along with comparisons with other cities. Together they give a general description of the distribution of land uses in the city.

Chief interest here is in commercial land use. Figure 11-3 displays in greater detail the location of all retail business groups and all retail stores which are isolated rather than within a group. Just as a comparison of the map of business locations with that of traffic flows clearly indicates locational relationships so does a relationship appear in a comparison of business locations with the distribution of residences and other land uses.

The short notes on various characteristics of traffic and land use in Cedar Rapids need not be elaborated, for in total they describe the study area adequately for our purposes and at the same time demonstrate that the area contains features of traffic and land use similar to those which are found in most urban communities.

Analysis of Retail Business Structure

Determining the exact pattern of the spatial arrangement of establishments is

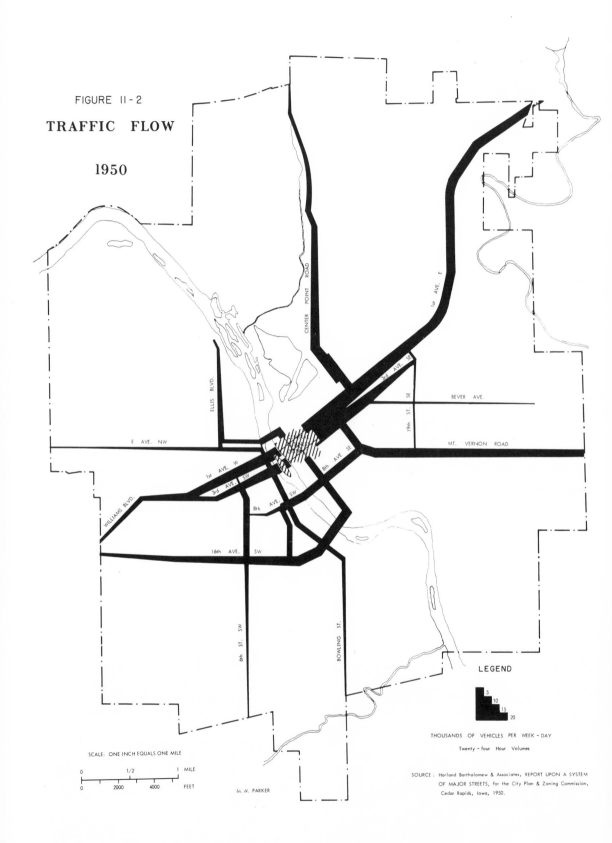

FIGURE II - 2

TRAFFIC FLOW

1950

LEGEND

THOUSANDS OF VEHICLES PER WEEK - DAY

Twenty - four Hour Volumes

SOURCE : Harland Bartholomew & Associates, REPORT UPON A SYSTEM
OF MAJOR STREETS, for the City Plan & Zoning Commission,
Cedar Rapids, Iowa, 1950.

SCALE: ONE INCH EQUALS ONE MILE

0   1/2   1 MILE
0   2000   4000   FEET

M. W. PARKER

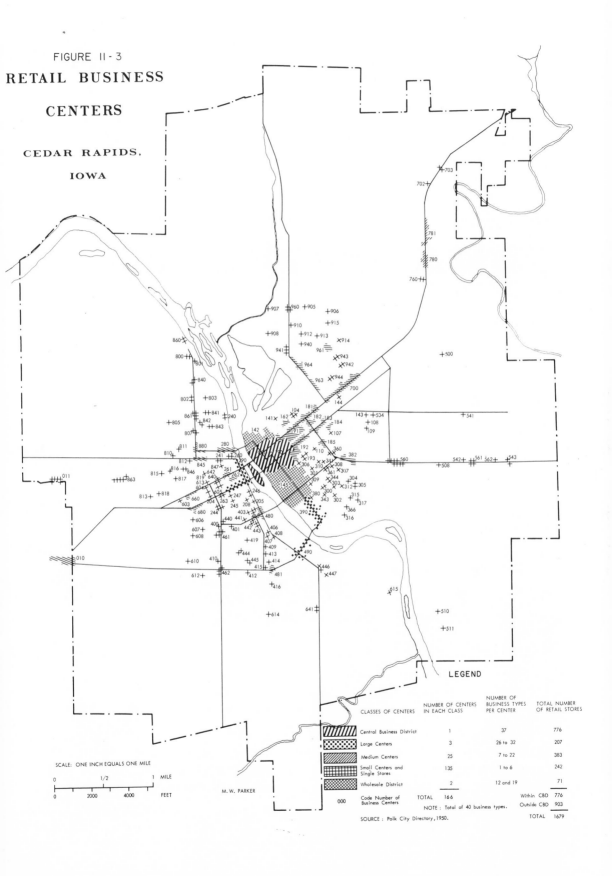

FIGURE II - 3

RETAIL BUSINESS

CENTERS

CEDAR RAPIDS,

IOWA

LEGEND

CLASSES OF CENTERS		NUMBER OF CENTERS IN EACH CLASS	NUMBER OF BUSINESS TYPES PER CENTER	TOTAL NUMBER OF RETAIL STORES
	Central Business District	1	37	776
	Large Centers	3	26 to 32	207
	Medium Centers	25	7 to 22	383
	Small Centers and Single Stores	135	1 to 6	242
	Wholesale District	2	12 and 19	71

000   Code Number of Business Centers   TOTAL   166

NOTE : Total of 40 business types.

SOURCE : Polk City Directory, 1950.

Within CBD   776
Outside CBD   903

TOTAL   1679

SCALE: ONE INCH EQUALS ONE MILE

0   1/2   1 MILE

0   2000   4000   FEET

M. W. PARKER

TABLE 11-IV
AREAS OCCUPIED BY EXISTING LAND USES, 1953
CEDAR RAPIDS, IOWA

Land Use	Area in Acres	Per Cent of Total Area		Per Cent of Total Developed Area	
		City of Cedar Rapids	Average in 13 Cities*	City of Cedar Rapids	Average in 13 Cities*
Single-family	3,483.7	19.19	18.17	39.98	30.92
Two-family	227.4	1.25	2.42	2.61	4.11
Multiple-family	118.8	.65	1.25	1.36	2.13
Commercial	267.6	1.47	1.52	3.07	2.58
Light industry	226.1	1.25	1.23	2.60	2.09
Heavy industry	232.8	1.28	1.59	2.67	2.70
Railroads	498.4	2.75	2.85	5.72	4.85
Public and semipublic	902.0	4.97	6.36	10.35	10.82
Parks and playgrounds	510.9	2.82	3.84	5.86	6.53
Streets and alleys	2,246.5	12.38	19.55	25.78	33.27
Total developed area	8,714.2	48.01	58.78	100.00	100.00
River and lakes	665.0	3.66	2.42		
Vacant land	8,771.8	48.33	38.80		
Total city area	18,151.0	100.00	100.00		

*Data on average of 13 cities of population 50,000-100,000.
Source: Bartholomew and Associates (1953).

essential to the analysis of business location in relationship to movement of customers. Since the movement data that will enter into the analysis later were collected in 1949, it is necessary to reconstruct the retail business structure as it existed in that year.

Collection of Data

The 1950 *Polk City Directory* (1949 data) was the primary source of data. A listing by address and kind of establishment of all retail business was compiled from the directory and these were plotted on a map of the city. As it is sometimes difficult to distinguish the function of businesses from the classes used in the directory, functions were checked by visiting sites. Many times the business was still in operation at the same address so its function could be determined. If the business was no longer at the address, conferences with persons familiar with the businesses of the city were sufficient to establish the activities which were carried on by that business.

Types of Business

The classification of business types was developed from the data. The businesses recognized and combined into forty types are listed below:

Grocery
Gas station
Restaurant
   drive-in restaurant
   root beer stand
Barber
Cleaner and laundry
   hat cleaner
Auto repair
   garage
   radiator service
   body and fender shop
   auto rebuild
   wrecking service
Drug store
   pharmacy
   sundries
Real estate, insurance, lawyer
   insurance (all types)
   loan company
   tax consultant
   law firm
Hardware and paint store
   garden equipment
Beauty salon
Tavern
   night club
Radio-TV sales and service
   electrical repair
Shoe repair
Clothing and shoe store
   secondhand clothing
   corsetière
Variety
Furniture
   antique shop
   bed and bedding
   home maker
   home decorator and furnishing
   auction
   secondhand furniture
Building service

roofer
carpenter
insulation service
heating installation
plumbing
Appliance
   gas appliance
   electric motor
   sewing machine
Bakery
   doughnut shop
Automobile dealer
   new car sales
   truck sales
Lumber yard
Doctor
   psychologist
Other professional services
   optometrist
   chiropractor
   podiatrist
   masseur or masseuse
   veterinarian
   small animal hospital
Department store or dry goods
Farm equipment and supplies
   windmill and pump
   feed and seed
Fix-it shop
   bicycle
   welding and sheet metal
   saw filing
   electric motor winding
   miscellaneous
   doll repair
Miscellaneous retail
   stationery, gift, hobby, and novelty
   leather store (harness and saddle)
   phonograph
   sporting
   rental
   photographer

music and dance studio
 jewelry and watch repair
Coal and fuel
 home oil
 ice and bottled water
Bank
Ice cream
 confectionary
 dairy
Funeral home
 monument
Hotel or motel
 trailer camp
Theater and other amusement
 billiard and bowling
 dance hall
 roller rink
 health studio and gym

Supermarket
Meat and vegetable market
 meat and fish
 vegetable and fruit
Automobile accessory
 tire
Used car
 junk dealer
Other building supplies
 brick and tile
 tent and awning
 floor covering
Dentist
Furniture repair and upholstery
 cabinet
 upholstery
 furniture repair

Spatial Associations

The purpose of the analysis of business arrangement is to determine if different retail types are commonly found in spatial association with one another. Another purpose is to determine whether or not these associations of functions are arranged in a hierarchy of business centers. Occurrence in the same shopping center is defined as an instance of association and the number of functions present in a center is a measure of its size.

Locations at both ends of the scale of size of center were not included in the analysis. First, business establishments at the homes of the owners were omitted. These include offices and studios of professional persons. Doctors' and dentists' offices at home are exceptions to this rule, however, and are included in the analysis.

The central business district is the other location left out of this part of the analysis. All business types used in the study, with the exception of lumber yards, funeral homes, and furniture repair shops, appear in the central business district. Many other retail businesses not found in any other center are also present in the central business district. The association of several hundred stores in this center overshadows associations in smaller centers. The central business district is a class by itself and is the central place of the hierarchy of businesses. Its inclusion in the analysis is not necessary and during the empirical analysis might obscure associations of businesses in other centers.

Shopping Center Defined

A shopping center is defined as an association of stores each one of which is within at least 200 feet of another retail store. When in the judgment of the observer modification of this general definition was warranted, more discriminating boundaries were established in the field. Most of the centers are easily defined. Two conditions where subjective judgment is required occasionally arose. Sometimes single grocery stores are within 200 feet of each other, one at each end of a short block in a residential area. The stores are considered separate centers in this instance.

Definition of functionally different areas near the central business district depends upon recognition of sudden changes in land use; this was the second instance where a subjective judgment was required. To the east and west of the central business district, wholesale houses and light industry become the dominant land use. The retail stores located within these areas as shown in Figure 11-4 are assigned to two centers, one on each side of the central business district, even though they may be farther apart than the defining distance.

Figure 11-4. On the left is a street scene of the major concentration of retail stores, the central business district. At the other extreme in retail locations are single standing stores such as the one shown on the right. Both are elements in the urban retail system of Cedar Rapids.

Automobile sales and used car lots are the dominant land use north from the central business district along 1st, 2nd, and 3rd avenues SE. A division between these areas and the central business district is along 6th Street SE at the line where an abrupt land use change occurs. The businesses are less than 200 feet apart, but an obvious change in the character of the district makes admissible classifying the area into separate centers.

One hundred sixty-five centers other than the central business district were established. Seventy-two of these were single, isolated stores. Stating which functions commonly appear together in these centers requires simultaneous observation of 40 geographical distributions, one for each type of retail business (see the list above). To achieve this, the study uses the same technique of analysis that was used in Chapter 4 of this volume. The measures of association employed are correlation coefficients calculated for all pairs of business types by using their occurrence together in a shopping center as the observations on which the correlations are made. Once correlation coefficients are assigned to each pair of business types, linkage analysis is employed to develop groups of business types in which every individual business type has at least one higher correlation with another business type within the group than with any type outside the group. The procedure used follows that of McQuitty (1957).

The analysis employed the 40 retail types observed in the 165 centers outside the central business district. A 40 by 40 matrix of correlation coefficients resulted. The matrix was reduced into ten *groups* of functions which were most alike under the definition of linkage analysis. These groups were:

1. Department store
   Bank
   Tavern
   Appliance
   Meat and vegetable
   Radio-TV sales and repair
   Furniture repair
   Funeral
2. Clothing and shoe store
   Other professional
3. Doctor
   Dentist
4. Auto dealer
   Used car
   Auto accessory
5. Heating and plumbing
   Cleaner and laundry
6. Supermarket
   Variety
7. Lumber
   Coal-fuel

8. Bakery
   Grocery
   Hardware and paint
   Drugs
   Beauty
9. Ice cream
   Confectionary
   Fix-it
   Barber
   Shoe repair
10. Restaurant
    Gas station
    Hotel and motel
    Auto repair
    Furniture
    Theater and other amusement
    Miscellaneous retail
    Building supplies
    Lawyer, real estate, loan and insurance
    Farm equipment and supplies

These ten groups were then associated with one another by finding the average correlation coefficient between each group and every other group. This additional grouping was undertaken to determine if two distinct higher order groups of business types could be found: one group associated with the arterials of the city and another associated with nucleated centers which are distributed more as the population of the city. Such a dichotomy was found by Berry in Spokane (Chapter 4). Although the business groups in Cedar Rapids did not display two higher order groups, it is interesting that groups four, seven, and ten, which most resemble the supplies-repair-arterial conformation found by Berry were highly associated. But group ten also linked with group nine, which was associated with the other groups, and this linkage prevented an arterial conformation from appearing. It is felt that these results neither prove nor disprove the existence of two higher order groups of businesses. The analytical procedures employed are quite crude and may not be able to distinguish differences which may be small in a city of the size of Cedar Rapids.

The next step in the analysis was to describe business centers of different sizes by the kinds of functions they normally contained. That is, a center with a given number of functions is described by the average number of stores of each function one finds in a center of that size. Correlation coefficients were again calculated and used as the measure of association. In this case, the size of the center is the variable and the average number of stores of each function observed in centers of that size are the observations. The correlation coefficients so determined were again grouped by linkage analysis.

In this manner three levels of centers outside the central business district were recognized by variations in the functions they contained. Thus a four-level hierarchy of business centers is established, the central business district being accepted as the highest order center.

Two centers displayed an unusual combination of retail businesses. They were the two wholesale districts and were retained as a separate group. The distribution of stores by groups of function and by level of center is given in Table 11-V.

TABLE 11-V
DISTRIBUTION OF STORES BY GROUPS OF BUSINESS TYPES
AND BY LEVEL OF CENTER

Groups of Business Types	Level of Center										
	Central Business district		Large Centers		Medium Centers		Small Centers		Wholesale		Total	
	Number	Per Cent	Number	Per Cent	Number	Per Cent	Number	Per Cent	Number	Per Cent	Number	Per Cent
Grocery, drug, hardware, beauty, bakery	43	16.6	23	8.8	84	32.1	111	42.5	1	0.3	262	100.0
Supermarket, variety	6	21.4	4	14.3	12	43.0	6	21.4			28	100.0
Cleaning-laundry, heating-plumbing	11	21.6	12	23.5	17	33.4	10	19.6	1	1.9	51	100.0
Shoe repair, ice cream bar, barber, fix-it, bicycle, welding	28	25.6	23	21.1	35	32.1	18	16.5	5	4.8	109	100.0
Auto dealer, used car, auto accessory	7	12.7	8	14.5	31	56.5	2	3.6	7	12.7	55	100.0
Lumber yard, coal-fuel dealer	3	7.3	2	4.9	13	31.6	7	17.0	16	39.2	41	100.0
Lawyer, real estate, insurance, * restaurant, theater & other amusement, misc. retail, auto repair, bldg. supply, furniture, farm equipment, gas station, hotel-motel†	409	55.8	60	8.2	140	19.3	75	10.4	38	5.3	722	100.0
Doctor, dentist	154	92.2	7	4.2	6	3.6					167	100.0
Clothing-shoe store, other professional	69	81.4	3	3.5	11	12.8	2	2.3			85	100.0
Department-dry goods store, ‡ bank, tavern, appliance, meat & vegetable, radio-TV, furniture repair, funeral	48	27.2	60	33.9	38	21.4	27	15.2	4	2.2	177	100.0
Other retail stores only found in CBD	52	100.0									52	100.0
Total retail stores	830	47.5	202	11.4	387	22.2	258	14.8	72	4.1	1,749	100.0
Total number of centers	1		3		25		135		2		166	

*Lawyer, real estate and insurance in CBD total 281.
†Hotel only in CBD and 1 in large center. Motels only in medium and small centers.
‡Department stores only in CBD.

Customer Movements

This description of customer movements is undertaken with the fact in mind that present location theory neither adequately explains why retail businesses should habitually cluster so tightly at the expense of paying higher rent nor does it reasonably define how customer travel motives and habits influence this clustering. As travel habits are certainly influenced by the capacity of the urban road network, very little understanding of the impact of highway change on businesses can be achieved by this approach until these questions of customer movement and business location are more clearly defined and understood. Thus an attempt is made to enrich location theory as well as to present descriptions of customer movements.

The assertion that customer travel is motivated by a desire to maximize travel benefits rather than to minimize travel costs (as implied by location theory) is advanced as the best available way to account for the travel characteristics described. Weighing the reasonableness of this assertion is one motive prompting the description of the particular elements of urban movement presented below. These elements are recognized in: (1) measures of the range of a good from each level of center and (2) measures of associations of purposes in trips. Customer movement from households to retail stores are described first. The size of the hinterland or extent of influence of a business type is suggested by this description. Measures of purposes combined on single trips are used to describe the business to business association established through the medium of customer travel.

Data

Concern in this study is with purposes of trips and the location of stops made to achieve these purposes. Data on these subjects were obtained by tabulating information on schedules prepared by respondents in the Cedar Rapids study area. For each trip purpose recognized the tabulation allowed cross-classifying: distance, location of terminal, frequency, and number and kinds of stops per trip. The sampling procedure and the characteristics of the respondents are described in Section III of this volume. The tabulation procedure involved tracing the routes traveled by individuals in the sample families and systematically recording where they stopped and how far they moved.

Trip purpose is defined by types of establishments visited. The exact address of each stop is given in the travel diaries. Any stop at a retail business is defined as the purpose of the stop regardless of the reason for entering the store. The one exception is when the person works at the store, in which case, the purpose of the stop was recorded as work regardless of the reason for entering the store.

Purpose of a stop has the same meaning as Mitchell and Rapkin's (1954) term *transaction* and implies actual behavior rather than behavior expected by the person as he sets out on a trip. The following purposes were recognized:

Forty types of purposes, one purpose for each of the *forty business types* recognized in the description of retail businesses.

Work and school purpose. These two trip destinations were combined into a single purpose because they display the same rhythmic pattern of trips.

Pay utility bill purpose. This purpose began to appear frequently once data processing had started and it was then included as a separate purpose. Gas, electric, and telephone company offices were terminals for this type of trip.

Other business purpose. This purpose includes: post office; business office; bus depot; freight and railway offices; and courthouse, unemployment office, and other government offices.

Other or *social-recreational purpose,* including stops at residences, churches, clubs, parks, race tracks, ball park, cemetery, Red Cross office, and offices of other institutions; also purposes which did not occur at establishments but were stated as purpose of trip, such as meeting friend and picking up or dropping off passenger, where the location of the action is given as street intersections rather than by establishments--all are included in this category.

General shopping purpose. Some respondents did not record exact address of stops as requested but just specified shopping by block location or corner. The exact purpose could not be established for such trips because most of them were in the central business district where many retail stores were in the immediate vicinity.

The locations visited in achieving trip purposes were divided in order to give specific attention to shopping trips. Detailed attention was given to shopping center locations and isolated establishments each of which is identified by a code number. Other locations were lumped into a single class. Thus all shopping trips were associated with at least one location with a code number. Work and school and social-recreational stops were either in a center and identified specifically or classed as "other."

The only trips used in the study were those which involved shopping as the purpose or one of the purposes. A trip by any member of a household was identified by the number and types of purposes it included. The identifications were divided in four categories: (1) single-purpose shopping, (2) multiple-purpose shopping, (3) shopping combined with work or school, and, (4) shopping combined with social-recreational purposes. These categories of trips were chosen because different travel patterns were expected to be associated with different travel motives. Greater travel costs measured by greater distance or frequency are likely to characterize trips where return for traveling is higher, for example, trips to work or trips including many shopping stops were expected to average longer distances than trips for single purposes.

The basic travel unit consisted of a round trip journey from home base to home base. Each individual in the family has a special role in its organization and therefore displays a special set of travel habits. For example, the child travels regularly to school, the head of the household to work, and the wife to shopping center or other locations. When two or more persons traveled to the same establishments the trip was considered a single family purpose. Thus joint trips to the theater, for example, were considered as one trip for that purpose and double counting avoided. Ambiguities arose when two persons in a single family made joint stops and other separate stops in the same trip. For example, if a father drops his daughter off at her place of work, then continues to his place of work, and later goes out to lunch, the stop at his daughter's place of work is recorded as in the "other" category of purposes, that is, dropping off a passenger. His trip purposes are "other," "work," and "restaurant." His daughter's trip will not appear if she goes directly home from work. If she stops to shop at the department store during her noon hour, her trip will be "work" and "department store." Thus two trips are recognized even though a segment of the trip was duplicated.

The round trip journey was traced from beginning to end and the length of each segment (distance between each stop) recorded. Trips were assigned a number and the household and person making the trip recorded. The following information for each household is available from this tabulation: (1) length and frequency of trips for every purpose, (2) the number and kinds of purposes associated in the same trip, and (3) the location of all stops.

Sample

The sample included 99 families living within the city limits of Cedar Rapids. All shopping trips made in a two-week period in May, 1949, by persons over 10 years old were recorded; a total of 1,160 shopping trips was made by this group during the period. The home locations of the respondents are shown in Figure 9-7.

Customer Travel: Households to Business Establishments

Presented here are combinations of trip data from the point of view of implied or stated theoretical statements on how people travel in urban areas. The first descriptions are summaries of movement from the household for shopping purposes.

Figure 11-5 presents frequency diagrams of the average distance traveled for the four categories of shopping trips. The number of trips which include combined purposes is greater than single-purpose trips, which indicates that multiple-purpose trips are an extremely important characteristic of household travel for retail goods.

It is clear from diagrams (b) through (e) in Figure 11-5 that opportunity for grouping trip purposes is important in allowing longer distances to be traveled per trip. From the point of view of the location of businesses, effective demand will be larger for an establishment spatially associated with other establishments (offering opportunity for multiple stops) than demand of his competitors not similarly located. The demand is larger because people are willing to travel farther to use the services of more than one store in the group. The kinds of associations appear important because of the different average distances involved in multiple-shopping trips compared with work trips and with social-recreational trips as seen in Figure 11-5.

A motivation for travel implied by combined-purpose trips is postulated as follows. Customers so acting are not attempting to minimize travel costs to obtain the services of one type of store. They may in fact go past a store offering the good that they obtain at a store with favorable associations. Rather they attempt to get maximum return from total travel effort needed to fulfill all the purposes combined on the trip. The effect on a business with a favorable association, be it with work place, social-recreational sites, or other business establishments, is expressed in the demand curve of the business, shifting it upward when compatible associations are found. Cost curves also will shift with location, rising because of increased rent in advantageous locations, but the business operator will not be concerned with rising costs if net gain rises a greater amount due to shifts in demand. Location theory clearly describes these adjustments of the firm.

It may be suggested that combined purposes on single trips reflect a similar motivation by customers. The opportunity for return (satisfaction) on a trip to some distant location is larger than the extra cost of reaching that location if trip purposes may be combined.

Figure 11-6 shows the distribution of the number of shopping trips per family during the period studied. The broad and uneven appearance indicates that travel

FIGURE 11-5
FREQUENCY DISTRIBUTIONS OF AVERAGE DISTANCE TRAVELED BY HOUSEHOLDS
BY VARIOUS CATEGORIES OF TRIPS

(a)

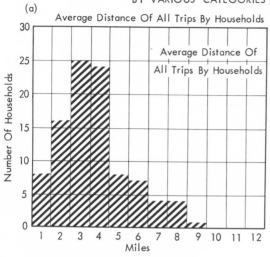

Note: Data refer to round trips
made by members of the
sample households
within Cedar Rapids city
limits.

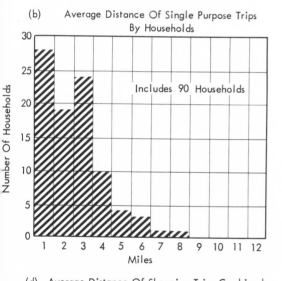

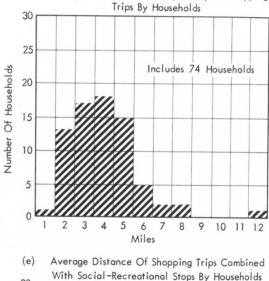

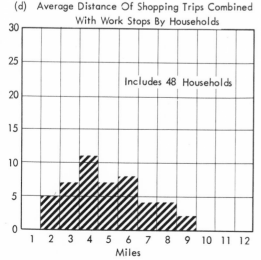

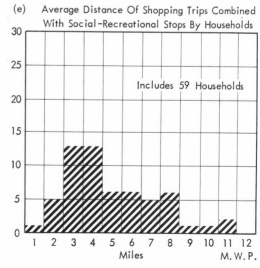

M. W. P.

FIGURE 11-6

FREQUENCY DISTRIBUTIONS OF AVERAGE NUMBER OF TRIPS PER WEEK BY HOUSEHOLD
AND BY VARIOUS CATEGORIES OF TRIPS

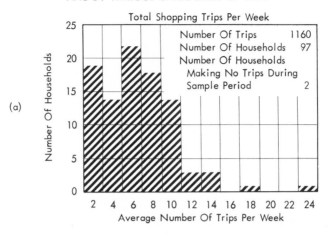

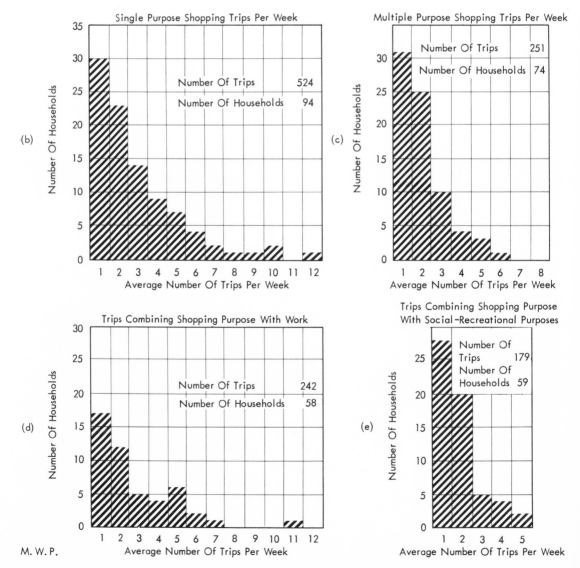

M. W. P.

FIGURE 11- 7

FREQUENCY DISTRIBUTION OF THE NUMBER OF CENTERS VISITED BY EACH HOUSEHOLD

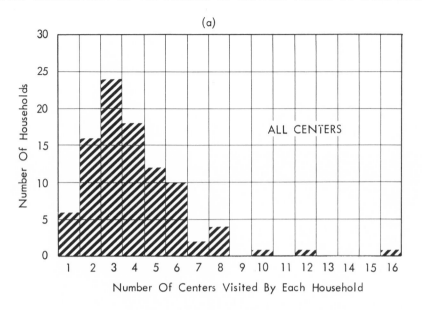

(a)

ALL CENTERS

Number Of Households

Number Of Centers Visited By Each Household

TRAVEL TO CERTAIN LEVELS OF SHOPPING CENTERS

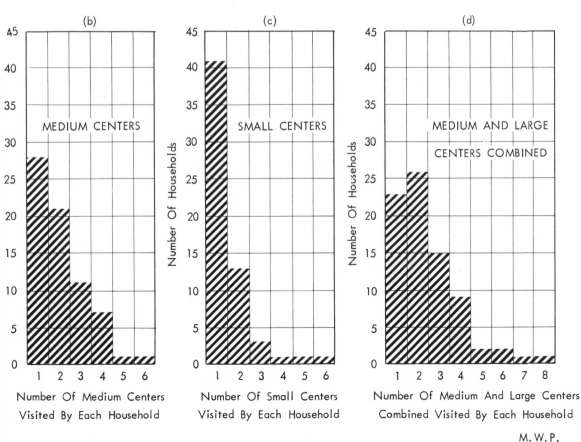

(b)

MEDIUM CENTERS

Number Of Medium Centers
Visited By Each Household

(c)

SMALL CENTERS

Number Of Households

Number Of Small Centers
Visited By Each Household

(d)

MEDIUM AND LARGE

CENTERS COMBINED

Number Of Households

Number Of Medium And Large Centers
Combined Visited By Each Household

M. W. P.

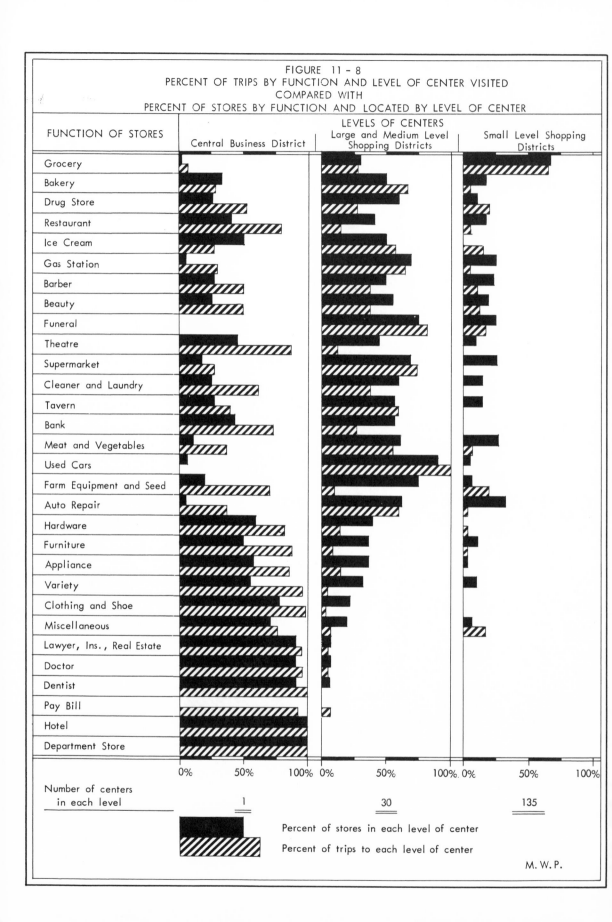

FIGURE 11 - 8
PERCENT OF TRIPS BY FUNCTION AND LEVEL OF CENTER VISITED
COMPARED WITH
PERCENT OF STORES BY FUNCTION AND LOCATED BY LEVEL OF CENTER

Percent of stores in each level of center

Percent of trips to each level of center

M. W. P.

frequency varies considerably from family to family. The variation in frequency is due to variations in travel requirements for families.

If a member of the family habitually makes a stop at a retail store such as a restaurant while on a work trip, orientation is to the place of work or perhaps along the route from work to home rather than to the home site. Such businesses may draw considerably farther than if standing alone, but the shape of the hinterland would be skewed by the location of major employment centers. It may be possible to abstract these employment centers to a few points such as the center of employment in the central business district and a few major manufacturing centers. Descriptions of hinterlands of suburban shopping centers shows they extend farther away from the central business district than towards it despite the fact population density is normally heavier towards the center. Perhaps a mechanism generating this pattern is an habitual stop at an establishment between home and work. In the aggregate, a dispersion area (consisting of home sites) would be described beyond an establishment located on an arterial leading outward from the central business district (considered as a major employment site).

The frequency of shopping trips combined with work and especially the variation in the curve at the five-trips-per-week level (corresponding to a five-day work week), shown in diagram (d) in Figure 11-6, suggest that shopping is regularly associated with work trips. The spatial adjustment suggested above may result from this travel habit. What proportion of shopping stops are in the vicinity of the work place or between work place and home requires further analysis not undertaken here. The importance of this trip characteristic is unknown but the implications are important when considering, for example, the effect of a limited access highway extending through an urban area. Would the area of dispersion be oriented around the exit points on the road? These suggestions are made only to demonstrate how causal relations can be postulated between the location of establishments, customer movement, and accessibility afforded by the road network.

Figure 11-7 displays travel habits to different shopping centers in the hierarchy. Centers are assigned to a level of center if the kinds and number of stores they contain resemble a center within the level more than any outside the level. Centers within a level of the hierarchy have approximately the same bundle of goods and services to offer.

Extremely important in this study is whether the shopping center hierarchy relates to the grouping of shopping purposes. The first column in Table 11-VI gives the per cent of families who visited at least one center in each level of the hier-

TABLE 11-VI

PER CENT OF HOUSEHOLDS VISITING EACH LEVEL OF CENTER AND PER CENT OF HOUSEHOLDS VISITING ONE OR MORE CENTERS IN EACH LEVEL

Level of Center	Per Cent of Households Visiting Each Level of Center	Number of Centers in Each Level	Percentage Distribution of Visits among Centers in Each Level					
			1 Center	2 Centers	3 Centers	4 Centers	5 Centers	6 Centers
Central business district	91.0	1	100					
Large centers	40.4	3	75.5	25.0				
Medium centers	69.8	25	40.6	30.5	15.9	10.2	1.4	1.4
Small centers	60.6	135	68.5	2.17	5.0	1.7	1.7	1.7
Wholesale centers	10.1	2	100					

archy. The next column lists the per cent of all families who make trips to that level but only to one center of that level. The other columns refer to the per cent who visit more than one center of that level.

According to the concept of a range of a good there would be no reason for families to patronize more than one center in a level (the nearest one) as all the goods offered by the level are found in that center. Considering that the hierarchy is poorly defined by the empirical procedures used in this study, a fair proportion of the trips by families is to only one center per level. However, the assertion that empirical procedures used in this study provide too rough a measure to identify discrepancy with theory can not hold in view of the large number of households which patronize more than one center of the same level. An enrichment of the notion of the range of a good seems to be required. One approach might treat the degree of imperfect knowledge on the part of the customer. Obviously customers travel to centers where they have some expectation of obtaining the services sought.

Perhaps expectations vary by type of good and level of center visited. A theory somewhat along the lines of that presented by Baumol and Ide (1956) in which travel motives are based on maximum probable net return is needed to explain customer travel habits under imperfect knowledge. A description of Baumol and Ide (1956) is presented in Section III. Statements of Troxel (1955), also reviewed in that section, are helpful as well.

Imperfect substitutability of products is also a factor motivating travel to more than one center for the same services. Better prices, wider quality range, or better association of goods and services for purposes of a certain trip may motivate travel to one level of center or another, or to centers in the same level.

An unexpected confirmation of the hierarchy of centers developed from a combination of trips to two levels of centers. The large level center consists of only three centers as shown in Figure 11-3. They are fairly close to the central business district and relatively few trips were made to them. It was felt that combining them with the medium centers would improve the sample by increasing the observations per cell. However, upon plotting the frequency of visits to each level as shown in Figure 11-7 [diagram (d)], combining the two levels resulted in visits to two centers being most frequent. It appears most households recognize enough difference between these levels of centers habitually to make trips to each level. This argues the merit of keeping the large and medium levels separated. The conclusion is based on number of trips and is independent of the criteria used in the procedure above for classifying the centers by group of functions present.

The graphs and tables just presented are summaries of trips by households. Another point of view is by type of good or service. Owing to the small size of the sample a detailed breakdown of trips into 41 retail business purposes by class of center or type of trip failed to fill each cell with an adequate number of trips. Business types which characteristically have few customers visit the establishment received very low numbers of trips in the sample.

Two business types, furniture repair and building services, received no trips at all. Ten other purposes with a total of less than ten trips each were dropped from the analysis because of the small number of sample trips. These purposes were: shoe repair; farm equipment and supplies; auto dealer (new); other building supply; other professional; lumber yard; coal-fuel dealer; fix-it, bicycle, and welding; building services; and funeral home.

Trips to each level of center were assembled by purpose of the trip (Table 11-VII). Despite the conclusion that households recognized a difference in large and medium centers (demonstrated by Figure 11-7d), when aggregating the data by business type rather than household so few entries (trips) were made to the large level center that the three large centers were included in the tabulation of medium level centers. Trips to the two wholesale centers were also added to this group for the same reason. Certain interesting trends are still discernible. The function of each level of the hierarchy is emphasized by the table because cells with less than five trips are not shown. In terms of location theory, a business type may be thought of as a bundle of goods and services. Each center contains a set of business types and offers a set of goods or services (for convenience called a set of goods). The higher centers offer the set of goods offered by the lower centers and in addition offer a variety of other commodities.

The average distance traveled to each level of center and for each good is listed in the table. The distance traveled for the low order goods increases as the class of center increases even though the high order centers tend to be nearer the center of the town and therefore closer together. Reading the table vertically shows that the distance traveled to obtain high order goods in high level centers is not significantly greater than the distance traveled to receive the low order goods from that level of center. These two tendencies (average travel to a level of center the same for all goods, and distance traveled for the same good increasing with level of center) imply that goods vary in range depending on the shopping complex; people travel through small centers where a good is available to obtain that good in a higher level center.

Customer Travel: Business Establishment to Business Establishment

The suggestion that people are willing to travel farther if they are able to combine trip purposes already has been made. Businesses which locate in places where many combined purposes may be accomplished may gain in competition with similar stores in other locations. The validity of the assertion depends upon the extent of combined purpose trips.

Out of 1,160 trips which included stops at retail stores, only 479, or 41 per cent, were for a single purpose, that is, a stop was made at only one store. Of these single-stop trips 201 were at groceries or supermarkets and 67 were at theaters. These three business types account for 55 per cent of all single-purpose trips. Nearly all other types were visited on trips where other stops were made as well.

Associations of retail businesses with other establishments through customers combining purposes on the same trip are common. As just noted, this occurs in more than half of the trips. How does combining purposes on the same trip affect travel for various goods? Table 11-VIII shows the average distance traveled on single-purpose trips and on multiple-purpose trips for stops at different types of stores (only types with a total of ten trips or more are given). The business types are ranked by the average distance of single-purpose trips. The low order and high order goods are again easily distinguished. The average distance traveled for a single-purpose trip is a measure of the range of influence of that type of store.

This distance is a round trip distance and may be thought of as the diameter rather than the radius of a circle around the store at the mean distance customers will

TABLE 11-VII

NUMBER OF TRIPS AND AVERAGE DISTANCE TRAVELED PER TRIP
TO LEVEL OF CENTER BY BUSINESS TYPE

Business Category	Central Business District		Large and Medium Centers Combined		Small Centers	
	Number of Trips	Average Distance Traveled in Miles	Number of Trips	Average Distance Traveled in Miles	Number of Trips	Average Distance Traveled in Miles
Grocery	15	2.9	76	1.3	173	.5
Bakery	11	3.8	17	1.5	6	.4
Drug store	87	4.2 3.7*	44	.9 1.5*	32	.8 1.2*
Restaurant	120	4.5	24	2.0	10	1.5
Ice cream	9	2.8	19	1.8	5	2.5
Gas station	18	4.4	40	1.4		
Barber	10	4.0	7	2.7		
Auto repair	11	3.0	17	2.2		
Beauty	8	3.7	6	1.0		
Funeral	0		5	1.7		
Theater	94	3.8	14	1.8		
Supermarket	41	4.3 4.0*	109	2.1 1.9*		
Cleaner, laundry	8	4.2	5	2.1		
Tavern	6	4.8	9	2.3		
Bank	43	4.6	17	2.1		
Meat and vegetable	10	3.6	16	1.6		
Used car	0		5	2.1		
Lawyer, insurance, real estate	17	2.9				
Hardware	22	3.7				
Clothing and shoe	63	4.8				
Variety	72	3.7				
Furniture	19	2.3				
Appliance	11	3.2 3.5*				
Doctor	47	2.8				
Department store	227	4.1				
Farm equipment, seed	7	4.4				
Miscellaneous	31	3.3				
Hotel	13	3.5				
Dentist	13	3.5				
Pay bill	32	3.7				

*Average distance for each group of functions by level of center.

travel for that store's service. It is road distance and not straight-line distance. Even considering these two facts, the mean distances given in Table 11-VIII show that considerable overlap of influence of a store in one center and a competitor in another center exists. The distance listed for each business type is much larger than the average distance between centers as shown in Figure 11-3 or the mean distance from the sample residences to the nearest center as listed in Table 9-III of Chapter 9.

TABLE 11-VIII

AVERAGE DISTANCE OF CUSTOMER TRAVEL TO BUSINESS TYPES;
SINGLE-PURPOSE TRIPS COMPARED TO MULTIPLE-PURPOSE TRIPS*
AND TOTAL TRIPS

Business Type	Number of Single-Purpose Trips	Average Distance	Number of Multiple-Purpose Trips	Average Distance	Total Trips	Average Distance
Grocery	157	.5	107	4.1	264	1.9
Beauty	4	.7	13	5.0	17	4.0
Bakery	5	.9	29	3.6	34	3.3
Drug store	57	1.1	103	4.1	160	3.0
Hardware	7	1.4	18	3.8	25	3.1
Restaurant	16	1.6	132	3.2	148	4.2
Barber	5	1.7	13	4.4	18	3.6
Appliance	2	1.7	10	5.6	12	4.9
Supermarket	44	1.8	100	3.4	144	2.9
Auto accessory	14	3.3	5	2.1	9	4.0
Bank	6	2.1	56	5.2	62	5.0
Meat, vegetable	5	2.2	19	3.2	24	3.0
Ice cream	13	2.2	20	5.6	33	4.3
Gas station	11	2.3	49	4.8	60	4.3
Pay bill	2	2.4	27	3.7	29	3.6
Tavern	4	2.4	12	5.2	16	4.8
Furniture	3	2.4	15	3.7	18	3.5
Dentist	6	2.4	7	3.4	13	3.0
Clothing	2	2.5	46	4.7	52	4.6
Lawyer	2	2.5	16	3.1	18	3.1
Auto repair	7	2.7	24	4.4	31	4.0
Theater	67	2.9	108	3.4	41	4.4
Department store	11	3.0	151	4.1	140	4.7
Hotel	4	3.1	9	3.4	13	3.3
Doctor	16	3.5	33	3.9	49	3.8
Miscellaneous	4	3.6	33	4.2	37	4.1
Variety	3	4.8	63	4.2	66	4.1

*Multiple-purpose trips include multiple-purpose shopping trips, trips with shopping combined with work, and trips with shopping purpose combined with social-recreation purposes.

Further analysis of single-purpose trips would be interesting but the small sample prevents making any conclusive statements. The average distance traveled to a type of store combined with the average frequency of visits would be a measure of the total use of the road network by a given type of store. Types of retail stores could then be ranked or classified by their use of the road network. Such a classification might be a first step in analyzing the impact of highway changes on different kinds of stores. It may be that a store whose customers do not travel far or who travel only infrequently is not affected by changed accessibility as much as a store where a great deal of customer traffic is required.

Because the differential sensitivity of business establishments to highway changes is important to questions of the impact of highway improvements (see the approach in Chapters 5 and 6), a list of establishments ranked by customer travel consumed is presented below. The ranking is from the largest to the smallest and the ranking is based on the aggregate of all establishments of a given type. The reader should keep in mind reservations regarding the size of the sample and the invariance of travel patterns under conditions of improved travel facilities (see Chapter 9) when examining the list. (* Indicates a tie.)

Department store	Furniture
Restaurant	Appliance
Grocery	*Insurance, real estate, lawyer
Drug store	*Laundry and dry cleaning
Supermarket	Farm equipment and supply
Theater	Auto accessory
Bank	Hotel and motel
Variety	Dentist
Gas station	*Funeral
Clothing and shoe	*Auto dealer (new)
Doctor	Shoe repair
Miscellaneous	Radio-TV sales and service
Ice cream store	Other professional
Auto repair	Other building supplies
Bakery	Used car
Utility company office	*Coal-fuel dealer
Hardware and paint	*Lumber yard
Meat and vegetable market	Fix-it, bicycle, and welding
*Beauty	*Building service, heating, plumb-
*Tavern	ing, etc.
Barber	*Furniture and upholstery repair

Table 11-IX is a summary indicating purposes which are generally found together. All the multiple-purpose trips were assembled by the kind of retail store visited. In order that comparison could be made a base of 100 was used. As a result Table 11-IX may be read: for every 100 trips to a business type there will be so many trips which include stops at work, so many which include stops at other businesses, and so on. Notice particularly in Table 11-IX business types such as restaurants

TABLE 11-IX
NUMBER OF STOPS AT OTHER ESTABLISHMENTS FOR ALL PURPOSES
(In Every 100 Trips by Business Types)

Business Types	Actual Number of Trips to Each Business Type	Factor Used to Adjust Number of Trips	Number of Single-Purpose Trips	Number of Stops for Work Purposes	Number of Stops for Social-Recreational Purposes	Number of Stops at Other Business Establishments
Department store	151	0.66	7	11	20	151
Clothing	52	1.92	4	8	15	253
Variety	66	1.52	5	20	14	275
Bank	60	1.67	8	48	20	190
Appliance	12	8.33	17	25	25	283
Miscellaneous, retail	37	2.70	11	24	46	208
Hardware and paint	25	8.00	28	4	8	188
Utility co. office	29	3.45	11	21	4	269
Doctor	49	2.04	33	10	6	112
Dentist	13	7.69	38	8	8	62
Hotel	12	8.33	41	8	33	108
Real estate, lawyer, insurance	13	5.56	11	11	22	195
Restaurant	145	0.69	11	52	26	122
Tavern	14	7.14	29	50	21	179
Theater	108	0.93	61	4	27	61
Beauty shop	16	6.25	31	6	38	131
Laundry	13	7.69	0	38	8	238
Barber	18	5.56	28	28	11	128
Auto accessory	14	7.14	64	7	0	36
Gas	60	1.67	18	33	45	65
Auto repair	30	3.33	23	33	13	97
Ice cream	32	3.13	41	0	38	68
Drug	160	0.63	36	8	23	113
Meat and vegetable	24	4.17	25	8	17	108
Bakery	33	3.03	15	15	24	142
Supermarket	144	0.69	34	12	19	106
Grocery	264	0.38	60	3	14	27
Furniture	18	5.56	17	17	11	245

that are associated with work sites. Also notice the high association with other businesses displayed by the central business district stores such as clothing, variety, and department stores.

Table 11-X gives a detailed breakdown of retail stores visited on single trips. Again, reading across, this table reveals how many times one would expect to see stops at one retail store type in every 100 trips to another store type. The blanks occur where at no time during the sample were the two retail purposes in that row and column combined. The asterisk occurs where eight or less trips in every 100 include stops at the other establishment.

The table vividly indicates the degree and range of associations of many business types.

Trip Purpose Groupings and Business Type Groupings

Attempts were made to group these data in a manner similar to the grouping of business types described earlier in this chapter. The result was a single group dominated by the department store type. A smaller grouping which included taverns, theater, beauty shop, and laundry tended to form around restaurants. Banks and

TABLE 11-X

NUMBER OF STOPS AT OTHER RETAIL ESTABLISHMENTS
IN EVERY ONE HUNDRED TRIPS TO EACH BUSINESS TYPE

Entries indicate the number of stops in 100 trips to the business types listed by row made to business types listed by columns.

Factor Used to Adjust Number of Trips to 100 per Business Type	Actual Number of Trips	Business Type	Number of Single Purpose Trips in 100 Trips per Business Type	Department Store	Clothing and Shoes	Variety	Banks	Appliances	Miscellaneous Retail	Hardware and Paint Retail	Utility Company Office	Doctor	Dentist	Hotel	Insurance, Real Estate, Lawyers	Restaurant	Tavern	Theatre	Beauty Shop	Barber	Laundry and Dry Cleaning	Auto Accessory	Gas Station	Auto Repair	Ice Cream Store	Drug Store	Meat & Vegetable Markets	Bakery	Supermarket	Grocery	Furniture
0.66	151	Department Store	7	50	22	11	*	*	*	9	*	*	*	*	13	*	*	*	*	*	*	*	*	*	19	19	*	19	*	*	*
1.92	52	Clothing and Shoes	4	94	25	17	*	*	*	*	*	*	*	*	21	*	*	*	*	*	*	*	*	*	12	12	*	*	*	*	*
1.52	66	Variety	5	91	15	17	*	17	*	*	*	*	*	*	15	*	*	*	*	*	*	*	*	*	17	17	*	15	9	*	*
1.67	60	Banks	8	40	17	10	*	*	10	*	*	*	*	*	38	*	*	*	*	*	*	*	*	*	*	33	*	*	*	*	*
8.33	12	Appliances	17	42	25	*	17	*	17	10	*	*	*	*	17	11	*	*	*	*	*	*	*	*	*	24	*	*	*	*	*
2.70	37	Miscellaneous Retail	11	43	22	*	*	11	*	*	12	*	*	*	14	11	*	*	*	*	*	*	*	*	*	16	14	*	*	*	*
4.00	25	Hardware and Paint	28	48	12	*	*	12	*	*	*	*	*	*	*	12	*	*	*	*	*	*	*	*	16	*	*	*	*	*
3.45	29	Utility Company Office	11	66	*	14	*	*	*	21	*	*	11	*	*	*	*	*	*	*	*	*	*	*	24	24	*	11	11	17	*
2.04	49	Doctor	33	39	*	*	*	*	*	*	*	*	*	*	*	*	*	*	*	*	*	*	*	*	15	24	*	*	*	*	*
7.69	13	Dentist	38	*	17	*	*	*	*	*	*	1	*	*	*	*	*	*	*	*	*	*	*	*	15	15	*	*	*	*	*
8.33	12	Hotel	41	*	*	*	*	*	*	*	*	*	33	33	*	*	*	*	*	*	*	*	*	*	11	*	*	*	*	*	*
5.56	18	Insurance, Real Estate, Lawyers	11	50	11	*	*	*	17	*	*	22	*	*	17	*	*	*	*	*	*	*	*	*	11	11	*	11	17	*	*
0.69	145	Restaurant	11	17	*	13	*	*	*	1	*	*	*	*	12	*	*	*	*	*	*	*	*	*	11	11	*	*	*	*	*
7.14	14	Tavern	29	*	14	21	*	*	*	*	*	*	*	50	*	*	*	*	*	*	*	*	*	*	21	21	*	*	*	*	*
0.93	108	Theatre	61	9	*	*	*	*	*	*	*	*	*	19	19	*	*	*	*	*	*	*	*	*	9	9	*	*	*	*	*
6.25	16	Beauty Shop	31	19	19	*	*	15	*	*	*	*	*	31	*	*	*	*	*	*	*	*	*	*	*	*	19	*	*	*	*
7.69	13	Laundry and Dry Cleaning	0	23	*	*	*	*	*	*	*	*	*	38	38	*	*	*	*	*	17	*	*	*	23	23	*	15	*	*	*
5.56	18	Barber	28	22	*	*	*	11	*	*	*	*	*	22	22	*	*	*	*	*	*	*	*	*	11	11	*	*	*	*	17
7.14	14	Auto Accessory	64	*	*	*	*	*	*	*	*	*	*	*	*	*	*	*	*	*	*	*	*	*	*	*	*	14	*	*	*
1.67	60	Gas Station	18	*	*	*	*	*	*	*	*	*	*	*	*	*	*	*	*	*	*	*	10	*	*	*	*	*	*	*	*
3.33	30	Auto Repair	23	10	*	*	*	*	*	*	*	*	*	*	*	*	*	*	*	*	*	*	10	*	*	*	*	*	10	*
3.13	32	Ice Cream Store	41	*	*	*	*	*	*	*	*	*	*	*	*	*	*	*	*	*	*	*	*	*	*	*	16	*	*	*	*
0.63	160	Drug Store	36	25	*	*	*	9	*	*	*	*	*	10	10	*	*	*	*	*	*	*	*	*	17	17	10	10	*	*	*
4.17	24	Meat & Vegetable Markets	25	*	*	*	*	*	13	13	*	*	*	*	*	*	*	*	*	*	*	*	*	*	17	17	21	17	*	*	*
3.03	33	Bakery	15	9	*	*	*	*	*	*	*	*	*	*	*	*	*	*	*	*	*	*	*	*	12	12	*	21	9	*	*
0.69	144	Supermarket	34	26	*	*	*	*	*	*	*	*	*	10	10	*	*	*	*	*	*	*	*	*	11	11	*	*	*	*	*
0.38	264	Grocery	60	*	*	*	*	*	*	*	*	*	*	*	*	*	*	*	*	*	*	*	*	*	*	*	18	*	*	*	*
5.56	18	Furniture	17	50	*	11	17	*	28	28	*	*	*	28	28	*	*	*	*	*	*	*	18	*	*	*	*	*	17	22

(*) indicates eight (8) or fewer stops were made at the business type listed in the column in every 100 trips to the business type listed in the row.

Blank cells indicate the business types were never associated in the same trip.

barber shops linked this group with the central department store group. Another group centering on grocery stores consisted of ice cream stores, auto accessory, and auto repair. Gas stations and furniture stores linked this group with the central group. One pair stood alone: hotels, and insurance-real estate-lawyer. These indistinct groups suggest the spatial grouping found in the previous analysis only in a general way. Establishing how influential the phenomenon of grouping is in the spatial ordering of businesses depends on development of more data and, perhaps, more sophisticated types of analysis.

Figure 11-8 compares the number of stores in a functional group at each level of center with the number of trips including stops at establishments in each group by level of center. There is a general agreement between proportion of stores and proportion of trips but with the proportion of trips normally larger in the larger centers. This may be an indication of increased scale of operations common to central business district businesses. The business types most highly associated with each level of center are apparent but, at the same time, the heavy overlap reveals a complex structure of business center and trips.

Although the central business district is but a single center, it has a large share of the business even of the low order goods (such as groceries which are available in almost every center). However, there does not seem to be an unusual degree of dominance by the central district in a city the size of Cedar Rapids.

Summary and Conclusions

The general discussion of Cedar Rapids presented at the beginning of this chapter is an attempt to make the study useful in the broad context of cities in general. Land use and traffic flow in Cedar Rapids are fairly typical of cities of comparable size, and it is reasonable to assume that the general arrangement of retail businesses and the movement of customers resemble the patterns in other cities.

The business arrangement which existed in 1949 in Cedar Rapids is described in detail and certain business types are found associated spatially in the outlying shopping centers. These groups are recognized as forming a hierarchy of business centers defined by the different sets of functions they offer.

Customer travel to business centers resembles the expected pattern but there are important exceptions with regard to distances traveled and association of stops. The average distance traveled for low order goods is much greater than expected. Persons travel through low level centers to high level centers to purchase items available in the low level centers. In certain other cases, where trip lengths are longer than expected (e. g., work, social-recreational, or other shopping), the combining of purposes seems to motivate the longer trips. An analysis of shopping stops on multipurpose trips shows a wide range of association which only generally resembles the spatial arrangement of business establishments. Thus grocery stores are weakly associated with other types both spatially and in single trips, whereas central business district business types, such as clothing and variety stores, are highly associated both spatially and in trip purposes.

The notions of expected patterns indicated in the paragraph above are notions from location theory. Since what is expected is incorrect, location theory needs to be modified.

Adopting the view of movement suggested by Troxel (1955) and Baumol and Ide

(1956) and other studies of consumer motivation and buying habits such as described in Section III appears to be a constructive way to modify location theory. This suggestion actually continues a line of criticism Lösch (1944) adopted in reviewing early attempts at location analysis. Lösch pointed out that early attempts were one-sided because they considered finding an optimum location for a firm or industry by analyzing cost factors whereas the true optimum was at the point of maximum net return. Using this more realistic point of view Lösch develops the network of central places. However, he uses as a criterion for establishing indifference boundaries around his market (or supply) areas the principle of minimization of travel costs. Isard (1956) formulates a similar system. These approaches are realistic when speaking of the optimum location of firms which have control over transportation costs of their raw material or products. When used in the context of explaining the arrangement of cities or shopping centers, the economic agent is no longer the firm but the customer who must decide how he is to travel to the firm. To maintain that he attempts to minimize his travel cost is one-sided in the same sense that concern only with minimization of cost is one-sided in the analysis of firm locations. Maximization of net travel returns is more likely the motivation for traveling in particular ways.

The results of this study do not make possible incisive statements on these points. Descriptions of the two distributions--spatial associations of businesses and associations of trip purposes--demonstrate both to be complex and their interaction even more complex. Massive data are required on an individual trip basis if business types which have few trips associated with them are to enter an analysis with a significant number of observations. Further analysis of the Cedar Rapids data is possible but other studies elsewhere of associations of purposes and locations at individual trip levels are desirable for it is in the characteristic bundle of trips associated with every kind of urban activity, of which customer travel to retail store is but one example, that the impact of highway change will manifest itself. From the point of view of this study these associations are the causal links between the urban road network and urban land uses.

SECTION V

Highways and Services: The Case of Physician Care

The task of this section is to present specific examples of the relationship between new highway construction, highway transportation, and the utilization of a service. Major effort is directed at the derivation and application of a model of the utilization of a service, which involves the characteristics of the transportation network as an integral element. The model measures not only the effects of highway construction on production, consumption, and the arrangement of trade areas, but also the benefits and losses from such construction to areas, producers, and consumers. This is undertaken in Chapter 14, "Models of Physician Utilization in Service Areas: Measurement of Benefits," using physician care as an example of service industry.

Chapters 12, "Highways, Service Areas, and the Example of Physician Care," and 13, "Empirical Studies of Physician Utilization," present the necessary background material for the analysis. The former includes an introductory discussion of the relationship of highways, service areas, and the movement of persons; a survey of pertinent market and trade theory; a discussion of highways and changes in medical practice; and a review of existing literature. The latter presents an analysis of the demand for physician care as well as empirical observations of movements of persons to physicians for Western Pennsylvania and Seattle. These observations are compared with theoretical results obtained through the model, and indication is given of the extent to which movement of persons may be predicted on the basis of present information.

12

Highways, Service Areas, and the Example of Physician Care

PHYSICIAN CARE is one example of the many services that involve primarily the movement of persons rather than goods. For reasons of manageability, it is used as the subject matter of this examination of certain geographic problems that arise in regard to any good or service. What forces govern the distribution of services and the differences in utilization of these services from place to place? In particular, what influences do changes in highways have upon the distribution and utilization of physicians, the size and shape of their service areas, and the movement of patients to physicians? The latter question is answered in three stages in the following three chapters. Clarification of the basis of the study demands that the concept of the service area be considered before proceeding to physician care in more detail.

Service Areas and Consumer Movement: Role of Highways

Like goods, services have a range and threshold market requirements (Berry and Garrison, 1958c and Chapter 3), that is, a minimum market is required for their support. This market is spatial. Any unit or aggregation of services must have a spatial market, more generally termed a service area, with local movement of customers within service areas to service centers and with interarea movement from smaller to larger service areas for the goods and services which smaller areas do not supply. Aspects of service areas are elaborated below.

The Basis for Movement or Trade

It is obvious that persons demanding services are separated in varying degrees from places supplying services. The distribution of places supplying services reflects the aggregate demand for a service but indivisibility and other economic characteristics of the supply activity produce the local spatial separation. A minimum aggregate or threshold of demand exists necessary to support a single supply unit such as a physician. Thresholds increase as the degree of specialization of the supply unit increases. Again, economic benefits are realized by producers in aggregates, a situation which requires larger areas of demand. Further, benefits also are derived by locating within recognized shopping districts so that multi-purpose trips are more easily satisfied. Therefore, a spatial arrangement is formed with the supply activity more concentrated than the persons demanding the service.

Characteristics of the transportation network have a major influence on these arrangements of services. Trade reflects the differential areal distribution of demand for and supply of a service but only in terms of possible routes between points of demand and supply. Routes have capacity and length, and are joined into a network. The transport network can be thought of as links and intersections imposed on an underlying pattern of demand and supply. This tends to channel supply and demand patterns in specific directions and volumes. But this is an oversimplification because the location of demanding and supplying units varies with the availability of transportation.

The Service Area

A service area is considered that area within which customers travel to a specific center and within which no alternate center can compete. In one empirical approach--the gravitational hypothesis--the boundaries between service areas are determined by a mathematical formula incorporating distance and the size of population of competing centers (Converse, 1949). This provides a useful but only an empirical approximation. From the point of view of theory, the boundary of an area is that line at which the sums of transportation costs to the competing centers added to the prices at the centers are equal. Prices at centers reflect supply and demand conditions within the areas and in addition will, of course, reflect the economic relationship of all areas (Enke, 1951).

If trade exists between centers for a given service available at both centers, then modern theory holds that a boundary between the areas cannot exist, since it is as cheap to move to the competing center as to shop locally. In general, trade implies the subservience of one place to another and its existence within the latter's service area. If areas are observed to overlap, then either the system is in disequilibrium, the location of the services is a matter of indifference over the whole area, or there are really two services.

In reality, particularly in the case of personal services, people do not always make decisions in strictly economic terms. There are overlappings of trade or service areas, long uneconomic movements, and subtle gradations in the attractiveness of supposedly homogeneous services. In the case of medical services, for example, people may go to the family doctor long after they have moved many miles away (Belcher, 1956). Also, people may go miles to a larger center on the expectation that the quality of medical care must be higher. These realities make it necessary to emphasize the noneconomic character of boundaries; people are generally unaware of fine economic distinctions, and economic solutions of service area problems are at best only approximations. Some of the data used in this study will help to indicate to what extent people are influenced by noneconomic values in their utilization of a particular service.

Systems of Service Areas

Services, therefore, operate in an interdeterminate system of service areas over space and are linked by the transport net.

The system of service areas is linked vertically as well as horizontally over space. In the medical services there are general practitioners, varying degrees of specialists, and hospitals, each with a different range, threshold, and size of service area. Although different in these ways, the common medical interest encourages aggregation and close association so that their systems of service areas are strongly related. For example, a hospital specialist medical service area of a city may coincide with and include a subset of general practitioner medical service areas.

Changes in Service Areas

Service areas are changed in three major ways: (1) by increases or decreases in the amount of service demanded or supplied, (2) by changes in the efficiency of the service, and (3) by changes in transport costs and transport networks. If the amount of services supplied increases faster than demand in a given area, the price will fall and the area expand at the expense of neighboring areas. It should be remarked that the distribution of urban centers from which particular services are supplied is a function of all goods and services. Hence, at any one time, imperfect spatial competition for particular services will exist, causing excess profits and capacity in some areas and losses in others (Berry and Garrison, 1958c, and Chapter 3 above). In general, new facilities tend to enter areas which offer a high income, that is, in which demand is increasing more rapidly than facilities and the price is high.

Of greater interest in this study is change in service areas induced by highway changes. An example of such a change is displayed in Figure 12-1. All peripheral areas have equal ease of access to the city with a surplus, city A; with highway improvements there would be reorientation of service areas. Service areas would be extended lengthwise along improved highways. Some might even be squeezed out of existence. Transport costs would be decreased. Hence, more transport would be consumed to take advantage of lower prices in A (see Chapter 1). In the absence of highway improvements, on the other hand, transportation costs increase gradually owing to congestion in A, which permits expansion of outlying centers.

FIGURE   12 - 1

CONFIGURATION   OF   SERVICE   AREAS

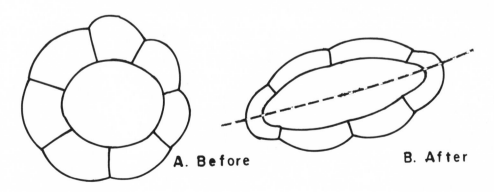

A. Before   B. After

Centralization and Decentralization Tendencies

In the past automobiles have been less prevalent and the ability to pay for transportation less widespread than today. Consequently, roads were in a primitive state, and people were unable to travel far for a service. Service facilities were more evenly distributed. A gradual rise in income has been accompanied by a demand for a higher quality of service, and technical improvements have increased the amount of specialization possible. These factors along with increase in the number of automobiles and improvement of highways have led to great concentrations of services in large cities and within cities in downtown districts (Nelson, 1955). But there have been certain counterfactors. As urban areas have grown and the downtown area has become more congested, there has been a rapid rise in transportation costs. There are, therefore, two contradictory tendencies: continuing increase in specialization and further concentration, as in downtown districts, versus decentralization owing to congestion. The increase of congestion and consequent rise in cost of transportation to reach the downtown core have also aided the rapid growth of significantly sized outlying shopping centers, permitting even the satisfaction of the threshold requirements for some specialist care (Weiskotten, 1952).

Several interesting problems present themselves in connection with changes in the transportation network and centralization and decentralization factors. If improvement in access to downtown areas is not made, a state of congestion results favoring the growth of decentralized centers. Congestion may be due to a lag during which most services are downtown while population decentralization has been extensive, or it may be due to sheer inadequacy of routes in relation to the traffic generated on the roads, or to both. We can profitably ask and perhaps determine whether the greater benefit to the region or urban area derives, in a given case, from enlarging the highway capacity to handle the demand more easily, regardless of the disparity of demand and supply of goods and services, or whether it would derive from an alternate pattern of services more decentralized. In short, are the benefits from a suggested highway improvement greater than an improvement in the structure of business? A second sort of problem arises when for other reasons, such as cross-city movement of through traffic, highway improvements are dictated. In this situation the very existence of a new facility will help define new patterns both of population and of demand and supply of services. The urban freeway is frequently of this sort, designed to improve downtown access and/or to provide a by-pass for through traffic.

The Interstate Highway System: Some Large Scale Effects

Construction of the Interstate Highway System may be expected to affect the size and pattern of service areas and the movements among them. The Interstate System links major national centers which are, of course, those with large amounts of services available. This availability reflects the general high income level of these areas, benefits of scale or association, and other large city advantages. Further, much of the construction on this system is to be within urban areas. Consequently, we might expect benefits to accrue to these large centers, which would increase the domination they enjoy and would magnify economic differences between large and small centers (Isard and Whitney, 1949).

This will occur because transportation is made cheaper; the service areas of the large centers will expand, particularly along the highways and at the expense of the smaller centers. The tendency toward concentration, already strong owing to increased specialization of services requiring large service area threshold populations, will be intensified. If within urban centers routes are adequate, they will tend to reverse urban dispersion which has been prevalent owing to the suburban movement of population and to congestion in central cities. It may be further concluded that centers on the system will have greater benefits than those not on the network, even if the general price of a service is reduced. This may appear as greater differentials between rural or small town income levels and urban levels. This income difference occurs since trade generally adds income to the exporting center and takes it from the smaller center. At the same time, however, the rural, small town areas may actually receive better services. On the other hand, improved access also appreciably raises urban rents (Isard, 1956), and this contradictory force may encourage decentralization of services to lower rent points on or near the Interstate System but outside of central cities.

Survey of Pertinent Market and Trade Theory

Given an area, a number of consumers and characteristics of their demand for goods and services, and technical characteristics of production and transportation, determine optimum patterns of settlement, production, market areas, and trade flows. This is the general location problem within which highway impact problems are imbedded. It includes, of course, factors of economic aggregation. The location problem is insoluble in its general form, but a number of special cases may be analyzed. Available special cases are trade and market area theory pertinent to the physicians' problem. For example, given the distribution and number of consumers, determine the location of production and identify market areas and trade flows. Or in another direction the problem of production and consumption may be reduced to a ranking of functions in order to determine the pattern of settlement. Finally, restricting the problem to given regions, one may ask what are their efficient patterns of production and trade? Although modern market area theory and trade theory are closely identified, early progress was in terms of one or the other.

Market area or service area theory is a restricted form of the location problem in which production and consumption locations are given and market areas are determined, usually by a simple weighting process (Converse, 1949) or by solution of the transportation problem of linear programing (Vidale, 1956). In some recent formulations attempts have been made at simultaneous determination of production levels, locations of supplying points, and market areas (Fisher, 1957). These recent formulations also use programing methods.

In its simplest form, trade theory treats two points and their flows of trade and traffic generating activities using the notion of comparative advantage. As with market area theory, this trade theory may be generalized and formulated as a programing problem. Formulations have many variations, and some of these involve the determination of both market areas and production locations (Brink and de Cani, 1957). One generalization is the spatial price equilibrium problem (Enke, 1951; Samuelson, 1952) which measures efficient production and consumption of regions as well as optimal routes and levels of trade. Generalization has been made to continuous space (Beckmann, 1953) which gives rise to determination of market boundaries and a continuous concept of trade.

Additional exposition is needed to clarify the character of market and trade theory; this is the subject of the following section. With reference to physicians' services, the task is to bring to bear the pertinent theory to the problem of measurement of changes and movements, including benefits and losses, resulting from changes in highways.

Relationships of Recent Theory

The pertinent theory can be grouped as follows:
1. Simple descriptive statements (e. g., gravity or weighting models in trade area work; simple statements of comparative advantages in trade analysis)
2A. Transportation problems (e. g., Vidale's formulation of the trade area problem, 1956)
2B. Spatial price equilibrium (e. g., Samuelson on trade, 1952)
2C. Economic aggregation (e. g., Fisher on trade areas, 1957)
3. Substitution analysis (e. g., Isard on the general location problem, 1956)

The simple descriptive approach is useful for the general formulation of problems but may not be used for the explicit solution of problems. Thus this approach is not treated further.

Transportation and spatial price equilibrium models attack the problem of trade and location in which locations of supplying points and receiving places, amounts to be supplied and amounts demanded, and transportation costs are given. The problem is to find least cost flows which meet the supply and demand conditions. The latter are more general in that amounts supplied or demanded are not set but are continuously variable with price. Both have restrictive cases, in which regions (locations) are given, and general cases, including one in which net trade is eliminated but markets, or supply areas, are determined. These are linear programming models and are generally solvable by analytic means. They normally have duals, which involve a measurement of values interpretable as benefits or losses. At present they are among the most valuable models in location theory and apply directly to the physician service example.

The economic aggregation model of Fisher (1957) allows variations in costs, or economics of scale, at supplying sites. Otherwise, it is quite similar to the models discussed above.

The general mathematical substitution analysis of Isard (1956) involves the use of transport inputs, which permits the stating of location theory in a manner similar to production theory and which admits many market points, many commodities, many producers, and many raw materials. The problem becomes one of complex substitution of all factors with profit maximization and cost minimization the objective.

Isard's formulation has the strong advantage of being very general. However, it seems most fruitful to use the spatial equilibrium model for the present problem.

The spatial equilibrium problem uses summary measures of admittedly more complex relations. On the other hand, it may be solved and its solution gives explicit measures of gains and losses of transportation changes. The detailed formulation of the model will be presented in Chapter 14.

The Case of Physicians' Services

It might be easily assumed that illness and the need for medical care and consequently the demand for medical services would reflect closely the distribution of the population. But demand for care and the distribution of physicians' services are not associated in close accordance with the distribution of population. In reality such factors as income, rural or urban residence, distance from a physician, age, and education--all influence the demand for medical care. As such, the demand for medical care depends upon the characteristics of areas and their populations and on the relative location of areas. Since the demand for care is channeled through the use of transportation services, the characteristics of transport media and carriers also influence the demand for medical care.

Nor does the demand for care coincide with the distribution of physicians and hospitals (Dickinson, 1954). The latter tend to be concentrated in the larger, older, and more prosperous cities. Here it is necessary to distinguish at least two medical services--that of the general practitioner who handles the large proportion of medical care that is more routine in nature, and the service of the specialist who handles cases in his particular field. As with classifications of many goods and services, there is no sharp break between the general practitioner and the specialist.

The distribution of general practitioners does not depart so radically from the distribution of population. But since people are willing to travel further for medical care than for most other services, the correspondence is not so close as for some services. As might be expected, specialists are heavily concentrated in larger metropolitan centers. People are willing to travel extraordinarily long distances to nationally famous clinics, for example. The distribution of general practitioners insures that most of the movements for medical care will be local, rarely passing over intervening physicians. For the care of specialists, on the other hand, there is a considerable regional and even interregional trade or movement.

Within urban areas, as in other regions, the distribution of physicians shows great concentration (Terris, 1956). In the case of cities, most physicians locate in medical centers in or adjoining the downtown district and in major shopping centers, but some general practitioners are distributed among lesser shopping centers and even in isolated clinics.

Distributions Studied

It was just noted that physicians' services have patterns of distribution that vary depending on the type of service. Also, features of the pattern that may be observed depend on whether the scale of view is large (say, the national pattern) or small (say, an urban pattern). In this study of physicians' services national, regional, and local distributions are subjected to analysis. Each level of generality leads to different but consistent conclusions. Variations in types of service are considered at each level of generality.

Some general understanding of the spatial aspects and character of medical care can be gained through a national analysis. It is possible to compare the known distribution of physicians with the estimated distribution of demand. An interregional model of trade in medical care (primarily for specialists) can be estimated and the relative importance of trade movements discovered. This serves as an example of the model of interregional trade and of the method for measurement of benefits.

Limited data are available for two cities and one small region: Seattle, Washington; Cedar Rapids, Iowa; and Western Pennsylvania, where some observations of movements as well as distances and routes traveled for medical care exist. In each of these areas it is possible to identify the pattern of movement and analyze relations between the distribution of demand, the business structure of medical services, and the interconnecting routes. In the first and last of the three areas, modifications in routes have either recently been completed or are envisioned for the near future. This provides the setting for the more general problem of simultaneously changing route structures, patterns of population and demand, physician location, and patterns of movement. The effect of route changes may be measured in terms of altered patterns of flow and of benefits or losses to areas and groups.

These studied distributions are only samples of one sector of the industry of the nation. The analysis of these samples cannot tell us the proper size and location of the proposed investment in highway facilities. The analysis can only serve as a rough indicator of what effects of highway development may be expected in other examples and in the aggregate.

Since highways and routes serve many purposes, at any one moment they can be taken as given for a particular industry. That is, the implication of the effect of route quality, capacity, length, and cost on the patterns both of population and of service facilities can be evaluated, but one should not proceed from an observed imbalance in one industry to a suggestion regarding route capacities and locations. However, the available problem is an important one. The inventory of services (as, for example, the number of physicians at a location and the size of service area this implies at any one time and place) depends for success on the ability of transport to bring demand and supply toward an equilibrium. The necessity to make decisions in planning business districts, business locations, and highways makes useful the understanding of how transportation in any one industry is demanded, satisfied, and changed.

Changing Medical Practice: Role of Highways

Stern (1945) and others have stressed how the last century has been one of most radical change in medical practice. This change has proceeded hand in hand with basic changes in our national economy--urbanization, specialization of activities, concentration of production, and increased income. All these changes are closely related to increased scientific and technical knowledge.

For the first 75 years of our national history, our country was essentially rural and rather sparsely settled. Medical practice was undifferentiated. The physician went to patients' homes, and time and cost of travel by horse and buggy were considerable. Consequently, doctors did not see many patients or serve a very large population or area. In the rural life, furthermore, demand for medical care was fairly uniform, and as a result doctors were distributed rather evenly about the countryside, in almost every village.

This pattern was fundamentally changed by three developments: the rise of cities and the differentiation of income; scientific discoveries; and the automobile. The rapid growth of cities created large and concentrated markets and new health problems as well. Gradually, the concept of specialization was introduced into medical schools. Specialists, of course, located in cities where a sufficient market for their more costly services could be assured.

Urbanization resulted in great differentiation in the attractive power of rural areas and of cities of varying sizes. The concentration of highest incomes or purchasing power in cities caused a rapid divergence in the ability of areas to compete for medical care. Very simply, most physicians were attracted to cities where incomes were higher, other amenities greater, and research and hospital facilities available.

The rapid increase in the number of automobiles and their widespread distribution has had a profound effect on the practice of medicine, both with respect to the distribution of physicians and in the relations of patients to physicians. Increasing numbers of automobiles accompanied the general rise in income in this century. Among the first users of the automobile was the doctor himself. He could see more patients a day--perhaps eight or ten rather than five or seven as before. Gradually, however, the rural population moved into towns to see the doctor, and often to live. Thus, owing partly to the automobile and partly to improvements of roads, patients now visit the doctors' offices. This in turn has greatly increased the productivity of physicians who are now able to see between 15 and 35 patients per day, having transferred the burden of transportation to the patient directly. Dickinson (1951) notes that in Pike County in rural Illinois, 16 physicians in 1950 provided more service to more people than 42 physicians in 1920. In the 1920's roads were so poor that physicians could take an hour or two getting five miles back to town from a rural call. It took six hours to get to the nearest hospital at Quincy, 40 miles away. Now patients come to town in 20 minutes, largely owing to the all-weather quality of rural roads, and Quincy is only an hour away.

Since people are willing to incur rather high transport costs for medical care, it is easy to see that this cost can approach the very cost of care. In the case of specialist care at regional medical centers in this country, demand for this care is so great that transportation costs may be higher than the cost of care itself.

Within cities the same change has occurred. As specialization developed, it was necessary to concentrate greatly in or close to the downtown area. Even before the widespread availability of automobiles a fair proportion of patients moved via mass transportation media. Most recently, greater use of the individual auto has increased congestion in the older medical districts. At the same time the population and income thresholds of outlying districts have permitted establishment of specialized care in these business districts. Again, these changes do not represent response to technical medical advance so much as to the complex interplay of population, automobile distribution, and the pattern of roads.

The general effect of automobiles has been to concentrate physicians in fewer settlements and to enlarge trading areas for medical service. In some cases rural areas have been left with inadequate local medical care, because of the greater attractiveness of cities, but often the areas are adequately supplied relative to the general care necessary. In any case it is impossible for rural areas to support

directly a wide range of specialized care. Obviously, this means that rural medical service areas will be deficient in total care, and alternatively, urban centers
are apt to have surpluses, which serve the surrounding region (Mott and Roemer,
(1948).

Existing Empirical Work

A large body of literature has been accumulated concerning medical economics.
The portion of the general field which is of concern here includes studies of the
demand for and the utilization of medical personnel and facilities. Data concerning
the characteristics of travel for medical care are included in some of the studies
of utilization.

Utilization of Medical Care

All studies emphasize that utilization of medical services is not primarily a
function of the prevalence of illness. Certainly most major illnesses are treated,
but the over-all utilization of a service varies in the extreme according to characteristics of the population and availability of the service.

Factors accounting for variability of utilization are above all income and the
relative supply of physicians, and lesser factors are distance, race, education,
and occupation (for example, Larson, 1952). The distribution of physicians (the
differential availability of services) itself reflects income and the "rural-urban"
and "size of place" characteristics of an area (Hoffer, 1950).

Distribution and Supply of Physicians

A great deal has been written on the subject of physician supply. Most studies
are not applicable because they consist of emotional and/or incomplete reactions
to the question of health insurance or the sufficiency of the number of doctors.
The prevalent social interpretation is that need is greater than supply and the disparity is becoming more acute. The economic interpretation is that supply matches
demand in the aggregate and that the difficulty lies in the distribution rather than
in the number of physicians (Dickinson, 1954).

The general distribution of physicians is given in Table 12-I (Dickinson, 1954).
The columns of population-physician ratios indicate variation in supply by regions
and emphasize the great departure of the physician distribution from the population distribution. Data by states, and especially by further disaggregations such
as Dickinson's "medical service areas," indicate even greater variation.

Physicians concentrate in the larger urban settlements. For example, the .7
per cent of places (107) over 100,000 population have 50.3 per cent of all physicians, and the 8.6 per cent of places over 10,000 population have 76.4 per cent
of all physicians. That the concentration is more extreme than that of the population can be seen in the fact that the same .7 per cent of places has 29.4 per cent
of the population and the 8.6 per cent over 10,000 population has 50.0 per cent.
Even within cities physicians are highly concentrated in certain districts. This
corroborates the economic interpretation of physician supply but does not of course
imply adequacy of care.

Studies indicate that the trend toward specialization has not abated. For example,

TABLE 12-I
REGIONAL DISTRIBUTION OF PHYSICIANS, 1950

Region	All Physicians	Population per Physician	General Practitioners	Population per General Practitioner	Full-Time Specialists	Population per F.T.S.	Other Physicians*	Population per Other Physician
Northeast	15,094	613	7,451	1,242	4,798	1,929	2,845	3,250
Middle Atlantic	50,929	586	25,049	1,191	16,895	1,766	8,985	3,321
South Atlantic	23,528	899	10,809	1,957	6,712	3,151	6,007	3,520
East N. Central	36,764	801	19,787	1,488	10,753	2,738	6,134	4,800
East S. Central	9,529	1,237	5,624	2,097	2,347	5,024	1,558	7,569
West N. Central	17,270	871	9,579	1,570	4,871	3,088	2,820	5,334
West S. Central	14,577	958	7,906	1,767	3,907	3,575	3,764	3,711
Mountain	5,977	840	3,076	1,633	1,736	2,893	1,165	4,311
Pacific Coast	20,904	686	10,138	1,415	6,784	2,115	3,982	3,603
UNITED STATES	194,572	770	99,419	1,502	58,803	2,548	36,350	4,123

*Other includes: interns, residents, and federal and state physicians.
Source: American Medical Association (1951) and Dickinson (1954).

Table 12-II demonstrates this trend in the graduates of medical schools (Weiskotten, 1952). The movement remains possible so long as population-income thresholds continue to be available in more and more places and the entire demand for specialists continues or increases. With respect to their distribution in the postwar period it is the smaller urban centers which are most rapidly supporting new

TABLE 12-II
PROPORTIONS OF GRADUATING CLASS ENTERING
VARIOUS OCCUPATIONS

	1915	1920	1925	1930	1935	1940	1945
General practitioner	22	24	25	32	25	23	19
Part-time specialist	36	40	41	38	18	13	6
Full-time specialist	41	35	34	30	56	64	74

Source: Weiskotten (1952).

specialists. The next table (12-III) gives the proportion of specialists for 1940 and 1950 in different size classes of cities. Also, considering all places in the United States, for places over 25,000 population specialists generally exceed general practitioners in number, but for communities under 25,000 the reverse is true.

TABLE 12-III
PROPORTION OF SPECIALISTS
BY SIZE OF PLACE

	1940	1950
Below 25,000	2.5	6.4
25,000-100,000	9.2	21.8
100,000-500,000	25.3	37.7
500,000-1,000,000	29.7	44.3
1,000,000 and over	29.1	45.1

Source: Weiskotten (1952).

Another set of studies traces changes in the distribution of physicians, especially the tendency toward concentration, over a period of time. The rate of concentration has been greater than the rate of urban growth and reflects technological change (specialization) as well as transportation improvement and increased rural-urban income differentials. An example is given in Table 12-IV (Nelson, 1955).

TABLE 12-IV
MINNESOTA: URBAN CONCENTRATION OF PHYSICIANS

City Population		1910	1930	1950	Per Cent Change		
					1910-30	1930-50	1910-50
Over 10,000	Physicians	1,050	1,992	3,022	89.7	51.7	187.8
	Population	685,000	997,200	1,260,000	45.5	26.3	83.9
2,500-10,000	Physicians	356	362	412	1.6	13.8	15.7
	Population	205,000	260,000	347,000	26.8	33.4	69.2
Under 2,500	Physicians	867	720	550	-16.9	-23.6	-36.5
	Population	1,225,000	1,307,000	1,402,000	6.6	7.2	14.4
STATE	Physicians	2,270	3,074	3,984	35.4	29.6	75.5
	Population	2,075,000	2,564,000	2,792,000	23.5	4.9	34.5

Source: Nelson (1955).

Note that only in the size class over 10,000 did the physician supply increase faster than the population. Such an extreme difference between this size class and others cannot be explained in terms of income or specialization alone, but demonstrates as well the effect of improved highway communication.

Studies of group practice have provided estimates of the demand for care in terms of numbers of physicians for a given aggregation of people (service area), as in Table 12-V.

TABLE 12-V
MINIMUM POPULATIONS FOR VARIOUS SERVICES

Service	Threshold Population
Physician	900 approx.
General practitioner	2,500*
Pediatrician	6,000
Internal medicine, surgery, eye and ear	10,000
Obstetrician	7,500
Ophthalmologist, radiologist	20,000
Urologist, orthopedist, dermatologist, and neurologist	40,000

* In small towns without specialist thresholds, general practice thresholds are smaller, 800 to 1,500 population.
Source: President's Committee on the Health Needs of the Nation (1952).

Another study of the range of services provided by all communities in an area considered the relative location of centers rather than absolute size, demonstrating, for example, the ability of the largest center to hinder the growth of a wide range of services in other nearby towns (Hassinger and McNamara, 1956).

Movement for Medical Care

Studies of movement for physician care are less common than those of the type just considered. Often medical trips are a minor part of a more general analysis such as origin-destination surveys.

Medical trips are among the longest for any purpose, since so much importance is attached to a particular doctor or quality of service and a great concentration of facilities does exist. Table 12-VI indicates the increase of length of trip as density, a crude measure of urbanization, decreases, and indicates as well the greater distance traveled to specialists, particularly in rural areas (Ciocco and Altman, 1954).

TABLE 12-VI
DISTANCES TRAVELED TO PHYSICIANS,
WESTERN PENNSYLVANIA

Area	Annual Visits per Capita			Population Density
	Total	General Practice	Specialists	
Allegheny	4.41	2.66	1.75	2,076
Lesser metropolitan centers	4.45	3.3	1.15	260
Adjacent to Allegheny	4.25	3.4	.85	250
Adjacent to lesser metro. centers	4.7	4.1	.6	75
Other	4.6	4.1	.5	65

Area	Average Distance (Miles) to General Practitioner			Average Distance (Miles) to Specialist		
	All	In County	Out County	All	In County	Out County
Allegheny	2.97	2.86	14.57	3.12	3.05	10.1
Lesser metropolitan centers	3.6	3.34	14.61	6.0	4.63	21.5
Adjacent to Allegheny	4.3	3.27	28.89	10.0	3.44	32.6
Adjacent to lesser metro. centers	5.3	3.92	20.13	14.1	5.10	38.0
Other	5.0	3.1	19.43	14.4	4.35	44.0
ALL	3.87	3.29	17.88	6.1	3.53	29.1

Note: "In County" and "Out County" refer to trips within or outside counties.

Source: Ciocco and Altman (1954).

Medical trips also differ from those for other goods or services in the degree to which they depend upon public transit and taxi. For example, whereas medical trips are about 1 per cent of trips in an urban area, medical trips are 2 per cent of transit trips and almost 8 per cent of taxi trips (Oregon State Highway Department, 1949).

Studies do not agree on the direct function of distance as a hindrance to the seeking of medical care. Apparently in high income areas with high car ownership ratios, distance is not a hindrance. However, in most situations the fact of distance and the cost it involves in time is a deterrent to seeking physician care (especially specialist care for rural residents). Table 12-VII indicates by distance zones from a physician the proportion of residents untreated or insufficiently treated (Hoffer, 1950).

TABLE 12-VII

PROPORTION OF RESIDENTS WITH SYMPTOMS UNTREATED OR WITH MINOR TREATMENT, BY DISTANCE ZONES

	Distance in Miles		
	0-5	6-10	Over 11
Some Treatment	24	29	37
Untreated	15	22	28

Source: Hoffer *et al.* (1950).

Some studies indicate a great overlap or absence of clearly defined service areas, especially where the local supply of physicians is inadequate (Belcher, 1956). A detailed study of outpatient and hospital visits indicates in general a strong movement from smaller to larger places for care, although a plurality may remain in the same size class (Table 12-VIII, Odoroff and Abbe, 1957a). In metropolitan areas the central cities care for a strong majority of the population of all zones. In fact, between 10 and 20 per cent of visits from nonmetropolitan regions have metropolitan areas as destinations. In addition, the survey of hospitals, the demand for whose services parallels the demand for specialists, has excellent data on distances traveled (Table 12-VIII). Distances are shortest in central cities and medium size cities and rise gradually through the urban fringe, smaller urban settlement areas, the rural nonfarm population in general, the rural nonfarm fringe of metropolitan areas, and the isolated rural population. In general the rural population travels about twice as far as the urban.

The most comprehensive study of movement and distances traveled is one undertaken in Western Pennsylvania (Ciocco and Altman, 1954). The study was based on a rather complete survey of physicians and identified intercounty movements, as well as distance traveled, for both general practitioners and specialists. Some of the data were given in Table 12-VI. More of the data from this study will be reviewed in Chapter 13. With regard to the role of distance, three statistically significant relationships were found: for general practitioners, distance traveled varied inversely with population density, which indicates that general practitioners are rather evenly distributed; for specialists, distance traveled rose as specialist

TABLE 12-VIII
A. GENERAL HOSPITAL USE BY RESIDENCE AND PLACE OF CARE
(Admissions per 1,000 Population)

Place of Residence	ALL	Metropolitan Area		Other Metro. Areas	Urban Areas		Rural Areas
		Central City	Outside Central City		10,000 to 50,000	Less than 10,000	
ALL	101	39	13	11	20	14	5
Metro. areas	97	68	21	5	2	1	
Central city	95	85	5	3	1	1	
Fringe	95	46	41	5	2		1
Rural nonfarm	107	59	33	10	4	2	
Rural farm	83	51	20	5	5	4	
Urban	115			18	61	34	2
10,000-50,000	117			18	93	4	2
Less than 10,000	113			18	23	69	3
Rural	100			18	34	30	18
Nonfarm	113			20	41	33	20
Farm	81			16	25	27	14

B. MEDIAN DISTANCE TRAVELED IN MILES

Place of Residence	ALL	Central City	Outside Central City	Other Metro. Areas	10,000 to 50,000	Less than 10,000	Rural Areas
ALL	7.4	6.2	6.1	40.1	7.7	8.6	8.6
Metro. areas	6.5	6.2	6.1	23.6	21.6		
Central city	5.6	5.4	6.4	55.0	9.0	16.9	
Fringe	6.6	6.8	5.8	77.3			
Rural nonfarm	10.1	12.5	6.6	17.9	17.5		
Rural farm	11.5	11.7					
Urban	6.5			37.2	5.8	5.8	15.7
10,000-50,000	5.9			18.4	5.3	11.9	
Less than 10,000	7.5			53.8	11.6	5.6	
Rural	13.3			47.2	10.7	12.7	8.2
Nonfarm	10.5			36.5	9.1	10.8	7.8
Farm	18.1			54.4	18.1	15.5	9.2

Source: Odoroff and Abbe'(1957a).

supply per population declined; for specialists, where distance to a specialist is greater, frequency of visits is lower (Table 12-VI).

The review of existing empirical work has served as a source of information in the determination of demand for medical care, as an indicator of the role of distance in the utilization of medical care, and as measure of the characteristics and changes in specialization and concentration of physicians. In later sections, data will be referred to as evidence for theoretical statements or for comparison with theoretically predicted results.

13

Empirical Studies of Physician Utilization

THIS CHAPTER reviews those empirical measures and descriptions which serve as the basis for the economic models of the subsequent chapter or as comparative statistics to be used as checks on the "goodness" of results. The more important of the empirical materials presented are the demand equations for physician care and the summary of interview data in Seattle.

The Demand for Physician Care

As indicated earlier, the demand for care in a large measure reflects economic variables rather than need for care. That is, demand may be expressed by an equation of the form, $f(p) = (x_1, \ldots, x_i)$, where the more significant independent variables are population and income; factors of occupation, race, and others which might also enter are highly intercorrelated with income and with each other. Further, there is little variation among regions in the relative importance of the variables. Hence a national demand equation is possible. The relationship bewteen demand and income, as indicated by data from the 1950 *Study of Consumer Expenditures* (University of Pennsylvania, 1954), is approximately linear. The following general equation was determined:

$$Lp = L(a + bI) - D \quad \text{or} \quad p = a + bI - \frac{D}{L}$$

where p = price in deviations from average national price
L = population (1950)
I = income (1949)
D = supply or demand for physicians, as in the transformation:

$$D = L(a + bI) \pm p(L)$$

which is to be interpreted as saying that an area's demand for physicians varies directly with population and income and inversely with positive price.

The other pertinent variable is transport cost as a function of distance. This enters in terms of price differentials. In reality, demand for physician care is rather inelastic so far as price differences due to local supply and demand conditions are concerned, but the friction of distance is real.

For all active physicians, the values of the variables are

$$D = L \,(.5518 + .0005601 \, I) \pm p(L)$$

so that the relative effect of population is 43 per cent and of income 57 per cent. Subequations can be expressed from the same data to distinguish between general practitioners and specialists. These subequations are

1. for specialists,

$$D = L \,(.14354 + .00018874 \, I) \pm p(L)$$

2. for general practitioners

$$D = L \,(.31802 + .00025839 \, I) \pm p(L)$$

As might be expected, demand for specialists depends to a far greater degree upon income than does demand for general practitioners. Later analysis indicates that there is error in grouping all specialists together, but the data did not permit finer distinctions. All coefficients were highly significant statistically, and the variables included account for about 80 per cent of the variation in observed demand.

These linear demand equations are a necessary part of an equilibrium model of utilization of physicians' services. Further, they have implications with regard to the supply of physicians. They describe a national equilibrium of supply and demand. Consequently, if any one area has an indicated deficit according to the equation, such a deficit cannot be made good efficiently by adding to the national supply. This would only make the surpluses of other areas greater, on the assumption that a present supply is sufficient. Therefore, additional physicians must be attracted to specific areas by providing the conditions equivalent to a greater demand in the equation (i.e., a subsidy of some sort).

The above equations assume that specialists and general practitioners are distinct entities and distinguishable one from another. The review of the literature (in which it was noted that people are willing to travel long distances for specialist care) helps establish the separate identity of the services. On the other hand, the degree of specialization varies widely, and at least for the less specialized end of the scale, a fair degree of substitutability exists (see also Table 12-VI). Not only are threshold requirements lower for the less specialized of the specialists, but often many general practitioners in rural or sparsely settled areas can satisfy the demand for these services. The reverse is also true. In metropolitan centers, specialists are substituted for general practitioners. If an equation is forced to express the demand for general practitioners for a large city, it may indicate a deficit in this service, but this merely reflects the use of specialists for more routine cases.

Physician Income and the Relative Supply of Physicians

It would be expected that income levels of physicians as between cities and regions would parallel that of the general population. This tendency certainly exists since expenditures per family generally increase with family income. Yet between specific areas and cities, the relationship is much less apparent. Physicians' incomes parallel the income of the population only if the supply of physicians closely matches the demand. If physicians are in relative excess, physician incomes will be relatively low; if they are in short supply, incomes will be high. The data in Table 13-I reflect both these tendencies and a number of exceptions (Weinfeld, 1951).

TABLE 13-I

PHYSICIAN INCOME AND THE SUPPLY OF PHYSICIANS

State	Physician Income	Relative Excess*	Per Capita Income	City	Physician Income	Relative Excess*
Michigan	$10,777	-190	$1,500	New York	$ 7,020	390
Washington	10,714	-110	1,520	Chicago	8,319	170
Minnesota	10,661	-40	1,190	Philadelphia	7,461	180
Idaho	10,375	-420	1,280	Boston	8,216	390
Oregon	10,375	-110	1,380	St. Louis	9,500	40
Arizona	10,333	-140	1,150	Cleveland	9,778	70
Texas	10,266	-210	1,260	Kansas City	11,500	-50
California	10,128	150	1,650	Baltimore	10,053	70
. . . .	. . . . .	. . . . .	. . . . .	Atlanta	10,450	-200
N. Carolina	8,526	-560	900	Louisville	11,308	-170
S. Carolina	8,319	-660	770	Los Angeles	8,624	200
Pennsylvania	8,106	30	1,100	San Francisco	10,100	200
Maine	7,736	-170	700	Minneapolis	9,888	10
New York	7,619	350	1,700	Milwaukee	9,727	0
Mississippi	7,586	-730	610	Seattle	10,667	150
Massachusetts	7,583	270	1,470	Portland	11,350	0
Rhode Island	7,282	0	1,470			
Arkansas	7,184	-410	770			

*Relative excess is per cent deviation index from mean U.S. population per physician.
Source: Weinfeld (1951).

It is noteworthy that the high income states of Massachusetts, New York, Rhode Island, Pennsylvania and the cities of New York, Philadelphia, Boston, Chicago, and Los Angeles have quite low physician incomes, and at the same time high ratios of physicians to population. The excess in specialists can be absorbed partly by competition for patients from outside the area, but the great excesses of general practitioners can be absorbed only by fewer patients and/or lower fees in these areas. On the other hand, states or cities with relative shortages of physicians-- Michigan, Idaho, Washington, Oregon, Arizona, Texas, Kansas City, Cleveland, Louisville, Seattle, Portland, and Milwaukee (often the areas which experienced great wartime growth) have the highest median physician incomes. Some states and cities with high incomes also have relative excesses (Minnesota, Baltimore, San Francisco, Minneapolis, and St. Louis). Each of these is a particularly important medical center, and apparently movement to them sufficiently increases competition for care to insure high income. Finally, in those areas with a low supply of physicians but also a low economic demand, physician income reflects general income (Arkansas, Mississippi, Maine, South Carolina, and North Carolina).

Information on physician income by size of community is summarized in Table 13-II. At the smallest size, in spite of great relative shortage, threshold market conditions do not obtain for a competitive income within the town size class of 2,500-5,000 population. At the other extreme, for cities over 500,000, physician incomes again decline, since effective demand fails to keep up with the supply of physicians. Apparently too many physicians are attracted to the "millionaire" class of city because of the general high income level as well as professional and cultural amenities. As a result, surpluses are great; incomes are forced down.

TABLE 13-II
PHYSICIAN INCOME AND SIZE OF COMMUNITY

Community Size	Excess*	Physicians per 100,000 people	Per Capita Income	
			of Physicians	of General Population
Less than 1,000	-73.5	29	$6,177	1,452
1,000-2,500			7,577	
2,500-5,000	8.5	120	9,050	1,913
5,000-10,000	8.5	120	9,354	
10,000-25,000	12.8	124	9,667	1,984
25,000-50,000	48.4	163	9,759	
50,000-100,000	39.1	154	9,793	2,128
100,000-250,000	76.2	195	9,537	
250,000-500,000	55.6	176	10,195	2,150
500,000-1,000,000	88.5	208	9,463	
1,000,000 and above	69.6	185	7,712	2,331
UNITED STATES	0	110	8,835	1,889

*Per cent above or below mean U.S. physicians per 100,000.
Source: Weinfeld (1951).

It is the medium size cities from 25,000 to 250,000 population which offer the greatest markets. Rural communities and small towns may have a strong need for care, but relative to the large towns and cities they cannot compete for the existing supply of physicians, so residents must travel to the larger cities.

Movement for Medical Care in Western Pennsylvania

This section reproduces and interprets the intercounty flow of persons for physician care in Western Pennsylvania (Ciocco and Altman, 1954). Other facets of this excellent study were treated earlier. Figure 13-1 identifies the patterns of intercounty flow. The most striking aspect of movement is the domination of Pittsburgh over adjacent counties. The existence of population thresholds for any and all services encourages a surplus aggregation which can compete over unusually

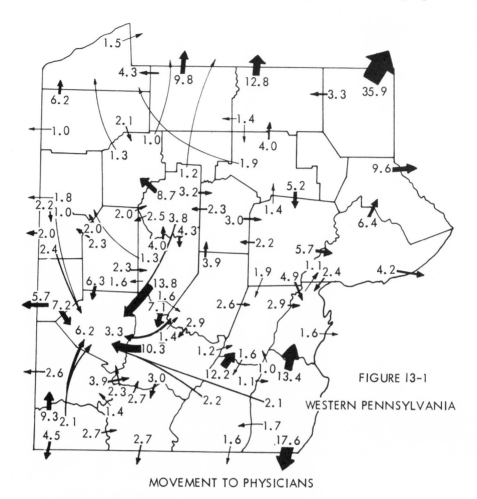

MOVEMENT TO PHYSICIANS

Figures represent proportion of patients leaving one county for another

long distances. This is due partly to the unique availability of certain services in Pittsburgh, partly to the fact that physicians, particularly specialists, prefer the assurance of part of a large market bloc, as long as patients are willing to substitute transportation cost for local convenience. In this way Pittsburgh's very existence hinders the development of middle-sized centers and tends to encourage long-distance travel and increased transportation.

Small flows are directed toward the lesser centers of Erie, Altoona, Johnstown, New Castle, and Sharon, and the out-of-area centers Youngstown, Morgantown, Cumberland, Williamsport, and others. In the main these are centers of specialist service areas. In the northest part of the region there is degeneration to local service areas and travel over great distances for higher specialities. No center is large enough to reach threshold levels for a service dominant over a large area. The very great number of local general practitioner service areas are not identified. The existence of many small intercounty flows is perhaps one indication of

the degree to which service areas do not coincide with counties. Small flows existing in both directions may indicate the lack of distinction of boundaries between
service areas.

Finally, note the distortion of the flow patterns due to the dominance of Pittsburgh. It should be noted that although Somerset and Bedford counties are tied to
Cambria and Blair, respectively, the accessibility to Pittsburgh afforded by the
turnpike does permit significant long-distance movement. In general, concentration
of an activity seems to be encouraged by economies of scale and good highway accessibility. A competing center depends upon distance-cost protection from the
price and variety advantages of the dominant center. In the northeast part again
this protection occurs locally, giving rise to many centers.

Changes in Location of Physicians: 1947-1957

An equilibrium tendency for income levels and patterns of physicians to coincide
with those of the general population has been noted. But population patterns are
dynamic. Are the locational patterns of physicians equally so? This question was
partially answered in two case studies, in Cedar Rapids and Seattle.

Cedar Rapids
Prior to 1950 Cedar Rapids was a member of that medium size class of city
that did not quite permit the establishment of large outlying business districts
owing to the absence of adequate thresholds (see Chapters 3 and 4 above). Ninety
per cent of physicians were concentrated in the downtown area. Since 1947 rapid
expansion has taken place in the urban fringe, particularly toward the north and
east (see Figure 13-2). A series of centers has developed located successively
further out on the major highway toward Marion. The downtown area has experienced an absolute decline of 31 per cent to a new proportion of 58 per cent of the
total number of physicians, while the outlying centers have grown fivefold. There
are very few isolated physicians. Physicians seem very dependent, as do other
businesses, on shopping center locations and locations on arterials. Although new
locations reflect the movement of the population, the greater use of the automobile
and the resultant parking and congestion problems downtown are also important
factors. To an extent, however, the close outlying centers are really new extensions of the Central Business District (CBD), perhaps the beginning of a medical
district.

Seattle
Seattle, of course, is a far larger city with an older development of community
shopping districts (often originally separate towns). This is particularly true north
of Lake Washington Ship Canal and Lake Union, which represent a partial barrier
to the greater dominance of the CBD in terms of ease of access. Well over one-half
of the non-CBD physicians were located north of the canal in 1947. Even so, fully
80 per cent of physicians in 1947 were located centrally: 58 per cent in downtown
office buildings, and 22 per cent in the First Hill hospital-clinic district. Seattle,
like Cedar Rapids, has experienced rapid suburban growth both of population and,

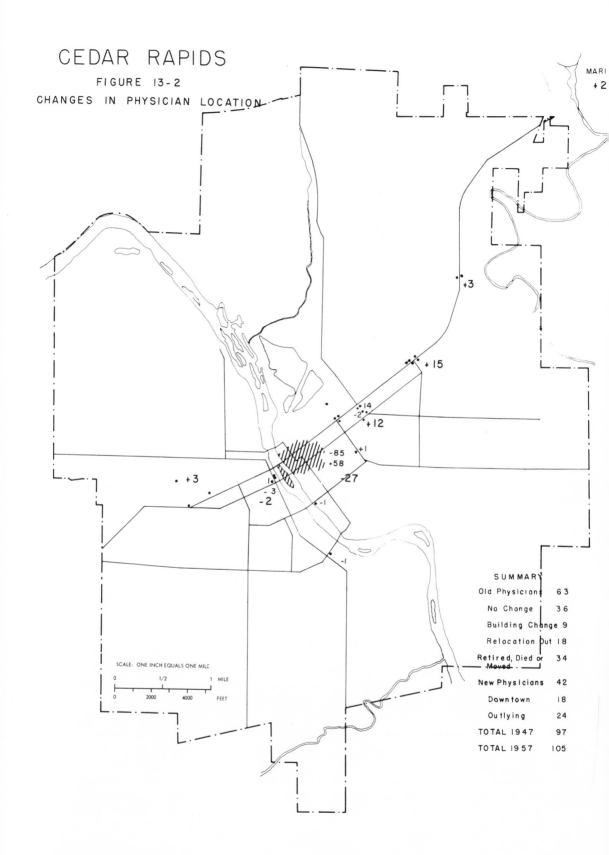

CEDAR RAPIDS
FIGURE 13-2
CHANGES IN PHYSICIAN LOCATION

MARI
+2

+3

+15

•14
-2
+12

-85
+58
+1

+3

-1
-3
-2

-27

-1

-1

SCALE: ONE INCH EQUALS ONE MILE

0   1/2   1 MILE

0   2000   4000   FEET

SUMMARY
Old Physicians   63
No Change   36
Building Change   9
Relocation Out   18
Retired, Died or   34
Moved

New Physicians   42
Downtown   18
Outlying   24
TOTAL 1947   97
TOTAL 1957   105

more strikingly, of shopping centers. Table 13-III indicates the changes in physician location by areas from 1947 to 1957.

TABLE 13-III
CHANGES IN PHYSICIAN LOCATION: SEATTLE 1947-1957

Area	1947	1957	Net Change	Same	Abandoned	Moved*	Per Cent Change	
							Absolute	Relative
Central								
CBD	418	401	-17	232	142	44	-4	-35
First Hill	163	288	125	104	39	20	77	20
Close in	26	80	54	13	4	9		
Suburban								
North	83	156	73	32	37	14		
East	1	32	31	1			156	74
South	37	108	71	15	13	9		
ALL	723	1,066	343	395	234	94	47.5	0

*Moved out of the area.
Sources: American Medical Association (1951 and 1957).

By 1957, the proportion of physicians in outlying business districts rose 75 per cent from 20 to 35 per cent of the total. In addition this table indicates a rather high rate of mobility and turnover in physicians. About 15 per cent of physicians moved outward along with the population, but almost a third retired or otherwise left practice in the decade.

Interestingly, the northern area, with the greatest growth in population, had the smallest relative increase in physicians. Possible explanations include delayed effects of improved access to the First Hill medical district and downtown via the Aurora Bridge, a possible smaller proportion of women and children, as well as the existent head start. On the other hand, to the east in Bellevue, demand for care is largely from women and children, much of which is most effectively met locally in spite of good highway access downtown. Particularly large developments have occurred at a considerable distance from downtown Seattle in Northgate, Renton, and Burien where distance hinders the attractiveness of the city center. The absolute decline of downtown locations and the growth of nearby First Hill, perhaps reflect extreme downtown congestion, and, in the medical district, the general growth of hospitals, clinics, and group practice, the ease of referral, the availability of laboratory facilities, and somewhat lower rents.

Seattle: Movement of Persons to Physicians, 1958

Data on trips to physicians were obtained from a sample of 35 physicians for a period of a week. The number of trips as well as other information are presented on maps, and the results are summarized in Table 13-IV. The primary value of the data lies in the description of the pattern of movement, which may then be

TABLE 13-IV
DISTRIBUTION OF TRIPS TO PHYSICIANS: SEATTLE, 1958

Area	Proportion of Income	Number of Trips to								TOTAL
		Burien		University		Northgate		Downtown		
		No.	Per Cent	No.	Per Cent	No.	Per Cent	No.	Per Cent	
North of city	7.37			6	14	16	22	54	6.3	76
East of city	6.56			1				43	5.0	44
South of city	11.52	58	88					46	5.3	104
Northgate-Lake City	9.35			3	7	32	45	63	7.3	98
University-Ballard	22.22			25	51	18	25	172	20.0	213
Close in	16.11			7	16	3		200	24.3	210
Downtown	3.16					2		59	6.9	61
Rainier Valley	9.42							88	10.3	88
West Seattle	12.10	8	12	2				73	8.5	83
TOTAL IN AREA	100.00	66	100	44	100	72	99	797	92.5	979
Out of area						1	1	65	7.5	66
TOTAL						73	100	862	100.0	1.045

compared with theoretical patterns. This helps show the extent to which persons operate within an economic equilibrium pattern of market areas.

The distribution of Seattle physicians is given in Figure 13-3. There is a great concentration of physicians downtown, almost 65 per cent. Beyond the obvious dominance of this center, there seem to be three levels of service: the large outlying centers, the neighborhoods, and the isolated clinics. The first two categories are dominated by specialists, the latter by general practitioners. The downtown area has a full range of specialists, the large centers the more common specialists. For both classes there are associated hospitals. Consequently, it is necessary to recognize at least four services if sense is to be made of medical service areas. The whole area and more is the service area for downtown, which we may call A center, and all other service areas nest within this highest order area. The outlying centers in turn, B centers, dominate neighborhood centers and local offices. A *pro forma* designation of service areas may be formed on the basis of services available at these various centers (Figure 13-4). This can be compared later with one derived from a theoretical model, based however on only one service. It is significant that the B level center occurs only at considerable distance from the CBD, protected by distance and terrain and supported by sufficient local markets-- i.e., east of Lake Washington and north of the ship canal. In fact the downtown area itself is the B center for a very large and populous area. From the point of view of population and income West Seattle ought to be a B center, but the fact that it lacks the unity of a distinct business center (perhaps owing to the string commercial development) may discourage medical service development. Apparently the hierarchical structure (Chapters 3 and 4) is very often broken in realistic situations, especially by the eclipse of the next lowest order center by the higher order center in its vicinity. This is permitted because economies of transportation and benefits of association permit the same area to provide different services.

Figure 13-5 compares the purchasing power of recognized districts of Seattle with

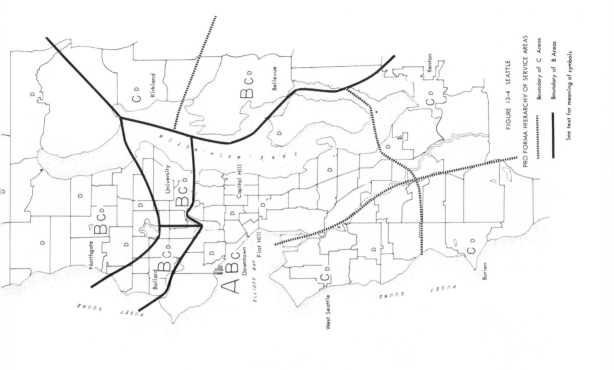

FIGURE 13-4 SEATTLE

PRO FORMA HIERARCHY OF SERVICE AREAS

━━━━━ Boundary of C Areas

············· Boundary of B Areas

━━━━━ See text for meaning of symbols

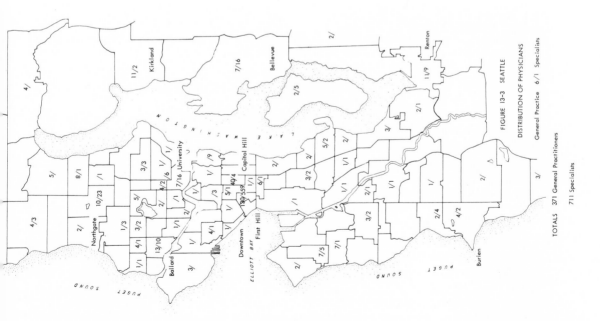

FIGURE 13-3 SEATTLE

DISTRIBUTION OF PHYSICIANS

General Practice 6/1 Specialists

TOTALS 371 General Practitioners

711 Specialists

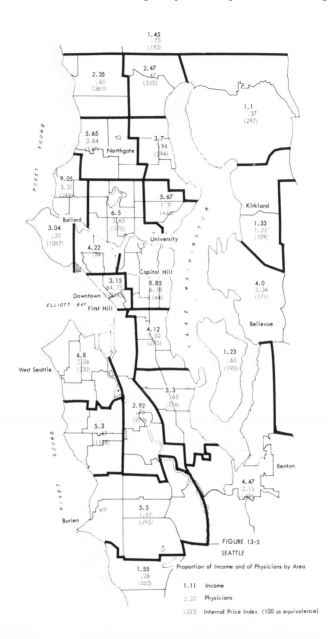

FIGURE 13-5
SEATTLE
Proportion of Income and of Physicians by Area

1.11   Income

2.22   Physicians

(333)   Internal Price Index (100 as equivalence)

the proportion of physicians in the same areas. All areas except downtown and First Hill have a lower proportion of physicians, but it is those areas with large shopping districts which show the least imbalance. This is given by an index where 100 indicates equivalence of purchasing power and physicians. It follows that the areas with highest indices will have the greatest incentive for patients to move out of the area. Likewise the downtown has an incentive to attract customers. But if a patient passes a lower index number center on the way, part of his need for less

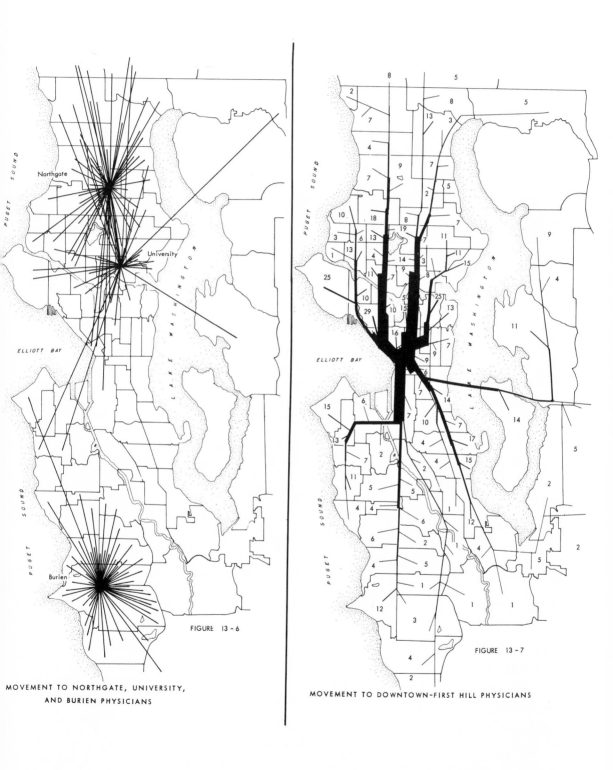

FIGURE 13-6

MOVEMENT TO NORTHGATE, UNIVERSITY,
AND BURIEN PHYSICIANS

FIGURE 13-7

MOVEMENT TO DOWNTOWN-FIRST HILL PHYSICIANS

specialized care may be satisfied there; for the most specialized care, however, even the largest outlying centers must look downtown.

Figures 13-6 and 13-7 show actual flows in terms of "desire lines" for Northgate, University District, and Burien physicians, and flows downtown and to First Hill. The Burien sample, a general practitioner, shows a restricted customer service area and little movement from beyond competing centers. The University District sample of specialists demonstrates a more erratic and widespread pattern of flow, which may be related to the large numbers of people who work in the area or attend the University. About one-third of the visits were from competing centers or beyond. For Northgate the pattern is as might be expected except for a few persons moving from downtown, which may be referrals. Movement downtown is placed on the likeliest roads, but the more significant information can be better seen from Table 13-IV. This table gives a breakdown by districts of flows to the sample physician areas and compares the proportion of flow to proportion of income. For example, the Burien area has 5 per cent of all income, but over 80 per cent of movement to Burien is from within the area. Similarly, for the University District and Northgate patients are much more concentrated near the physician location.

It is important to know whether or not the downtown area draws equally from all areas. If it does not, this is an indication of successful competition from outlying centers and indicates, therefore, the friction of distance which permitted the centers to develop. The data support our expectation, since all the areas close to downtown have higher than proportional response. Nevertheless, there are many general practitioners downtown who supply the lowest level of service to distant patients. This results partly from the general concentration of daytime population downtown, partly from the multipurpose possibility of trips. But the data indicate that patients of general practitioners are more centrally concentrated than those of specialists. Between one-fourth and one-fifth of all patients to downtown general practitioners were from distant areas, while one-third to one-half of patients to downtown specialists were from distant areas.

14

Models of Physician Utilization in Service Areas: Measurements of Benefits

THIS CHAPTER contains estimations of flows of patients and resulting patterns of service areas in contrast to the observed flows and areal patterns described in the preceding chapter. The estimates involve before and after solutions with reference to anticipated construction of the Interstate Highway System. This permits the estimation of the impact of highway development on the utilization of physicians' services.

The model involved is a modified transportation one, more specifically a spatial price-equilibrium model (Enke, 1951, Samuelson, 1952). A set of areas is assumed (for example, states, census tracts, or any arbitrary areas for which trade information is desired). The demand equation for physicians is then applied to each region, which yields an internal price that matches supply and demand. If the difference between the internal prices of two regions is greater than the transport cost between them, trade will take place unless usurped by another region with a greater difference. Trade will raise the price of the exporting region and lower that of the importing region so that at equilibrium the net prices will differ by the amount of the transport cost between them. This extends to the equilibrium for a system of service areas.

In the present analysis prices and costs have a somewhat special meaning to be explained later in the discussion.

The Primal Formulation

A formulation of the problem is given below. The primal solution is one of transport cost minimization relative to the optimal pattern of production, consumption, and trade (Fox and Taeuber, 1955). In order to display the problem, the equations are set down in a list below:

A demand equation(s) for a good or service is given by

$$p_i = f(x_1^* \ldots x_m^*, q_i) \text{ and } A_i = f(x_1^* \ldots x_m^*, z_i^*)$$

where p_i is equilibrium price, in the ith region, A_1 is internal price, * denotes predetermined variables, z_i is original internal supply, and q_i equilibrium supply (consumption).

The total supply or production of a good or service is given:

$$z_i^* = k$$

Gross equilibrium conditions occur when total supply equals total demand:

$$\sum q_i = \sum z_i^*$$

$i = 1$ to n $i = 1$ to n

A transport cost function is defined as:

$T_{ij} = f\,(M_{ij},\ Y_i \ldots Y_m)$ where M_{ij} is distance from region i to j.

Trade restrictions are:

$$p_i \leq p_j + T_{ji}$$

Trade may exist:

$$E_{ji} \geq 0 \text{ when } p_i = p_j + T_{ji}$$

Trade is impermissible:

$$E_{ji} = 0 \text{ when } p_i < p_j + T_{ji}$$

Exports equal imports:

$$E_{ji} = E_{ij}$$

A Dual Formulation: Measurements of Value or Benefits

The equation system for the measurement of benefits follows.

Costs of transportation, or the "value of the service," if positive, are a gain or positive benefit to transport services but a loss or indirect cost to the system; if negative, they may be considered as vehicular benefits.

$$\alpha T_{ij} \alpha E_{ij} = \alpha C_{ij}$$

It is very important to note that if there is no trade among areas there is no change and consequently no new benefit. Also, even if there is a net loss to the system, there might be a small per-vehicle or per-trip gain.

For Exporting Regions

The gain or loss to local producers or consumers is given by:

$$q_i p_i - z_i A_i = \alpha V_{ic}$$

This is a nonvehicular benefit.

Producers' gain due to increased trade or gross regional benefit is:

$$\alpha E_{ij}\,(p_j - A_i) = \alpha V_{it}$$

This is a nonvehicular and a reorganization benefit.

Total producers' gain is:

$$\alpha V_{ic} + \alpha V_{it} = \alpha V_{itc}$$

Net regional benefits are:

$\alpha V_{it} = \alpha V_i$, unless transport costs are charged to the producer. Then:

$$\alpha V_{it} - \alpha C_{ij} = \alpha V_i$$

For Importing Regions

The gain or loss to local consumers is given by:

$$z_j A_j - z_j p_j = \alpha V_{jc}$$

This is a nonvehicular benefit.

The gain or loss to local producers is given by:

$$q_j p_j - z_j A_j = \alpha V_{jp}$$

Consumers' loss due to trade is:

$$e_j p_j - e_j (A_j - p_j) = \alpha V_{jt}$$

This is a nonvehicular and a reorganization benefit and represents income leaving the area.

Total consumers' loss is given by:

$$\alpha V_{jc} + \alpha V_{jt} = \alpha V_{jtc}$$

Net regional benefits are:

$$\alpha V_{jtc} + \alpha V_{jp} \ (- C_{ij} \text{ if applicable}) = \alpha V_j$$

For the Entire System

System producers' gain or loss is:

$$\sum \alpha V_{itc} + \sum \alpha V_{jp} = \alpha V_p$$

System consumers' gain or loss is:

$$\sum \alpha V_{ic} + \sum \alpha V_{jtc} = \alpha V_c$$

Net social pay-off is given by any one of:

A $\quad \alpha V_p - \alpha V_c = \alpha V$

B $\quad \sum \alpha V_i - \sum \alpha V_j = \alpha V$

or it can be computed directly:

C $\quad \sum E_{ij} [(A_i - p_i) + (A_j - p_j) + t_{ij}]$

The symbols used in this formulation are the same as those in the primal formulation given previously. α refers to an incremental value.

The basis for measurement of benefits is from a maximization problem which accompanies the cost minimization problem. Benefits are a function of the exports (E_{ij}), internal and equilibrium prices (A's and p's) and internal and equilibrium supply and demand (z's and q's). Benefits are not necessarily realized in increased profits or savings.

Benefits are a measure of the net social pay-off of a new efficient arrangement, produced by some change, over a former efficient arrangement. In this case benefits are a measure of the value of more efficient transportation and of increased trade (both vehicular and nonvehicular benefits). Reorganization benefits accrue to individual areas but disappear in the aggregate since they measure the effects of altered patterns (Chapter 2). Negative figures may be interpreted as direct costs.

The social interpretation of benefits may be complex. Either consumers or producers may benefit most from a change; this depends partly on the allocation of transport costs. Producer benefits may be in the form of increased sales, and consumer benefits in the form of lower prices. However, benefits may be more indirect. Producers may be more fully employed; consumers may save through lower transport costs or through the avoidance of excess consumption. Highway improvement will lead to increased travel. However, increased transportation expenditures will be more than offset by benefits of a lower per-trip cost and greater volumes of trade.

Modifications and Additions to the Basic Model

Generally, spatial price equilibrium has been used just as a model of trade, but it is also used here for the purpose of estimation of benefits and of service areas. That is, the patterns of trade of given arbitrary regions will be found. Then service areas will be expanded or contracted until no trade exists, thus satisfying the theoretical conditions for a service area. Service areas can be contracted in the direction trade moves and expanded on the flank toward which trade comes. This will not be done in a highly rigorous fashion but as an approximation in this study. Finally, the solution after Interstate Highway construction will be imposed and the effects on service areas seen, as well as benefits and losses estimated.

Further desirable modifications which may be incorporated in a future study envisage taking into account economies of scale, threshold conditions for various services, and substitutability of services. These involve greater mathematical complexity but will better enable estimation, for example, of efficient changes in service location resulting from highway construction and other changes, rather than just the effects, the subject of this study.

Transport Costs

Transport costs, or the costs of movement of persons or goods, are a measure of the effect of geographic separation and its economic representation. Transport cost is rarely a simple function of distance. Either actual charges can be tabulated (Morrill, 1957) or a set of functions representing, for example, high, intermediate, and low rates per mile can be used (Fox and Taeuber, 1955). Influencing factors are the quality of transportation, such as the condition of the highways or economies of scale and of the long haul, and discriminatory rate structures. For purposes of costs of medical trips in this study distance is modified as to road quality, and a trip terminal charge is recognized. Transport costs do not appear as a determinate variable in the demand question, but since transport costs enter into the restrictions of regional price differentials in the programing formulation they strongly modify consumption of services in areas.

Unfortunately, little work has been done on the weighting of highway transport costs. It is necessary to accept time-cost as a measurable criterion of the friction of distance. Studies indicate between a twofold and a threefold variation in time-cost per unit of distance, for example, from a weight of .6 to 1.6 where .6

represents a limited access freeway, 1.0 a quality two-lane paved highway or urban signaled arterial, and 1.6 a dirt or heavily congested road (see, for example, Saal, 1955). The above weighting system is used as a rough but necessary modification of transport cost functions.

Changes incident to the construction of new highways are felt through a change in distance weights, and these changes alter the relative competitive position of areas. A standard form of transport equation is:

$$T_{ij} = a + bkM_{ij}$$

where a includes a fixed terminal charge, k a road quality weight. This transport cost is not necessarily of the same magnitude as actual cost since regional price differentials may not be as great as interregional transportation charges. The equations $p_i = p_j + T_{ji}$ must often have a different meaning from that normally used in spatial equilibrium theory. Since movements do occur, people must attach relatively less significance to transport costs than their actual computed cost. Hence, the term T_{ij} is not a measure of the computed cost of transport, but the cost as recognized by drivers. This, of course, involves estimation of the effect of non-economic values on economic variables (here costs).

Solutions

The primal solution of trade, production, and consumption is obtained through simple but tedious iterative techniques which alter the pattern of movement and the level of prices until the demand-supply equilibrium condition is achieved while satisfying other restrictions. The dual measurement of benefits (social pay-offs) is then a matter of arithmetic.

Meaning of Prices and Costs

It cannot be too strongly stated that both transport costs and price levels in this analysis are only approximations. Further analysis is required to bring the model into clearer relationship with actual figures and the parts of the model itself to equal precision. In this study prices are entered as deviations from a U.S. average of 0. For convenience, and since it would be pointless to be exact here but not elsewhere, the unit expenditure (U.S. average) per physician is assumed to be $100 (U.S. actual average expenditure per capita, 1949, $40). With reference to the demand equation this means that price is somewhat inelastic to demand changes: for example, a variation of 20 per cent in supply is accompanied by an 8 per cent change in price. This is relatively close to the little empirical information available (Ferber, 1955).

Transport costs, as noted earlier, are treated generally as having an impact equal to some fraction of the actual two-way cost of movement. The greater the specialty of care sought, the less significant is a given unit of distance. For a lower level of service which is generally available the relative role of transport costs increases. The levels used in this study were chosen in direct reflection of empirical observations of the distances and frequencies of actual movements but are not rooted to some theoretical function.

Consequent to these assumptions, cost or benefit figures in dollars can only be taken as indicating the approximate level of benefit or cost subject to a substantial

error, although the relative distribution of the·benefit as between regions or consumers and producers can be taken with more assurance.

A National Model

A regional distribution of physicians in the United States has already been given
(Table 13-I). It is relatively easy to solve the interregional equilibrium for the
nine-region case, but this is far too gross for purposes of estimating the effects
of highway changes incident to large scale highway developments. Nevertheless,
it may serve as an illustration of the method of allocating benefits. Two examples
are given. For 1950 the benefits of having interregional trade are computed; next
a solution is found which indicates the effects of changes in the distribution of physicians over time on the patterns of movement for care. Both examples use data
from the 1951 and 1957 American Medical Association Directories. A future study
envisages a many-region case for the United States which will specifically consider
the effects of the new highway construction.

Equilibrium Solution

The solution of the nine-region case for specialists in 1950 is presented in a map
(Figure 14-1). Included are net regional consumption, equilibrium prices, and
movements. Movements which are actually of patients are analyzed for convenience
in terms of their physician equivalents. Table 14-I presents as well the pre-equilibrium prices, initial supplies of physicians, and the demand for physicians for
1956 as well as 1950. The specific procedures of the solution, as well as the table
of transport costs, will not be given here. Examples of the techniques are available
(Fox and Taeuber, 1955; Morrill, 1957).

In the 1950 case largely because of the grossness of the regions, only those
paths of movement with greatest initial pre-equilibrium price differentials are used
at equilibrium. The other regions remain self-sufficient. For 1956, only two paths
of movement are permitted since, except for the East North Central and Middle
Atlantic regions, price differentials were reduced by 1956. Likewise, where these
differences were greater movements were correspondingly larger.

With reference to the theory of service areas or regions, it is clear that the influence and drawing power of the Atlantic Coast medical centers extends further
inland than the boundaries as arbitrarily chosen. That the distribution of physicians
is not at equilibrium is also easily seen by the existence of a movement of patients
from the East North Central to the New England states *across* the Middle Atlantic
states. It could be shown mathematically that a very great saving and benefit would
derive to these three regions if the distribution of physicians were more even. Such
a redistribution might take place in the long run, but the relative immobility of
physicians indicates the validity of measuring the benefits of changed patterns of
movement. The present imbalance is a result of many factors, including historical
lag, greater medical, educational, and research concentration in seaboard cities,
and a generally greater concentration of service activities.

Summary of Benefits: Nine-Region Case 1950

The benefits as derived by this solution are, for the purposes of the equation,

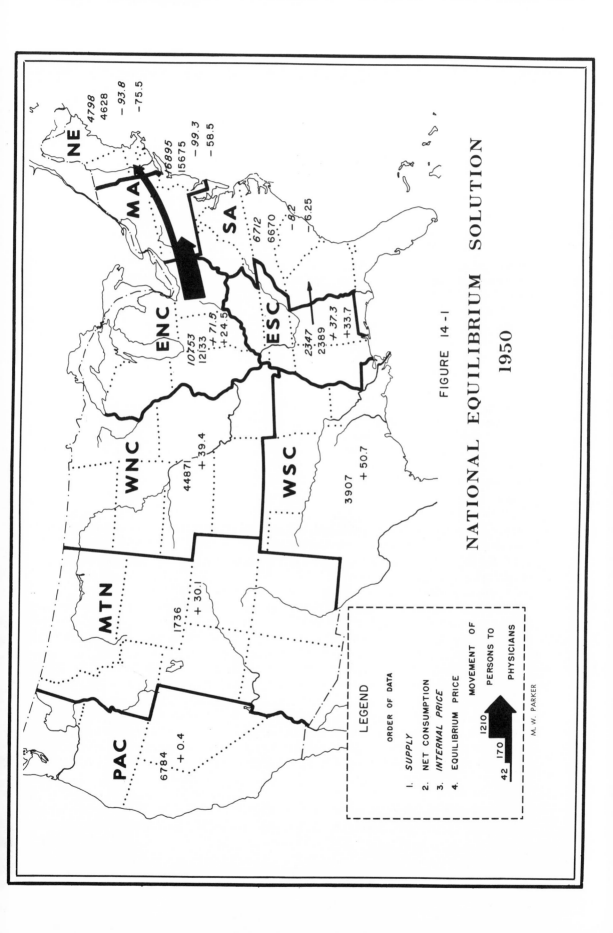

NATIONAL EQUILIBRIUM SOLUTION

1950

FIGURE 14-1

NE
4798
4628
− 93.8
− 75.5

MA
16895
15675
− 99.3
− 58.5

SA
6712
6670
− 8.2
− 6.25

ENC
10753
12133
+ 71.5
+ 24.5

ESC
2347
2389
+ 37.3
+ 33.7

WNC
4487l
+ 39.4

WSC
3907
+ 50.7

MTN
1736
+ 30.1

PAC
6784
+ 0.4

LEGEND

ORDER OF DATA

1. SUPPLY
2. NET CONSUMPTION
3. INTERNAL PRICE
4. EQUILIBRIUM PRICE

MOVEMENT OF

1210

PERSONS TO

PHYSICIANS

42   170

M. W. PARKER

TABLE 14-I
NATIONAL EQUILIBRIUM SOLUTION
A. 1950

Area	Initial Supply	Demand	Internal Price	Equilibrium Price	Net Consumption	Movement (Exports)			
						N.E.	M.A.	S.A.	TOTAL
N.E.	4,798	3,929	-93.8	-75.5	4,628				
M.A.	16,895	13,935	-99.3	-58.5	15,675				
S.A.	6,712	6,538	-8.2	-6.2	6,670				
E.N.C.	10,753	12,856	71.5	24.5	12,133	170	1,210		1,380
E.S.C.	2,347	2,787	37.3	33.7	2,389			42	42
W.N.C.	4,871	5,472	39.4	39.4	4,871				
W.S.C.	3,907	4,615	50.7	50.7	3,907				
M.	1,736	1,887	30.1	30.1	1,736				
P.	6,784	6,788	.4	.4	6,784				
U.S.	58,803	58,803	0	0	58,803	-170	-1,210	-42	± 1,422

B. 1956

Area	Initial Supply	Demand	Internal Price	Equilibrium Price	Net Consumption	Movement (Exports)			
						N.E.	M.A.	S.A.	TOTAL
N.E.	5,556	4,715	-87.5	-56.5	5,246				
M.A.	20,679	17,754	-101.5	-39.5	18,659				
S.A.	8,239	8,215	1.1	1.1	8,239				
E.N.C.	12,630	16,423	113.0	43.5	14,930	310	2,020		2,330
E.S.C.	3,011	3,161	12.8	12.8	3,011				
W.N.C.	5,536	5,721	12.5	12.5	5,536				
W.S.C.	5,060	5,325	16.9	16.9	5,060				
M.	2,419	2,253	-27.9	-27.9	2,419				
P.	9,195	8,961	-13.6	-13.6	9,195				
U.S.	71,302	71,302	0	0	71,302	-310	-2,020		± 2,330

in terms of $100 units of a physician's income. To make a very gross approxima-
tion to actual values, we may multiply the derived benefit by 88.3, the number of
$100 physician-income units in 1949. The results of the solution for benefits is
summarized in Table 14-II.

This breakdown indicates the distribution among regions and producers and consumers of the benefits of having trade over not having trade. Net consumer gains
or losses are negligible for all areas, but producer losses or gains are significant.
These are primarily nonvehicular and reorganization benefits. The present case,
of course, is meaningless in terms of an actual public improvement since it is hard
to imagine interregional movement as not existing. However, it is theoretically
significant in describing the benefits of trading and, if restrictions on international
movement are present, it has pertinence to national planning and international
trade policy. The meaning of the benefits is easily understood here. For example,
for the Middle Atlantic and New England states producers' prices rise as movement is possible and physicians become more fully employed. The magnitude of
their gain depends on the volume of movement. Consumers benefit, too, by not being encouraged by low prices to exceed needed levels of service. On the other hand,
for the East North Central states, producers incur indirect costs through loss of
a large captive market and lower prices, while consumers "lose" in spite of lower
prices, in the sense that expenditures rise, although medical care may be more

adequate. This means a significant cost or loss to the deficit regions since the income of the service consumed after movement accrues to the surplus region. Here is an example of a reorganization benefit. This has important meaning with reference to possible taxation. It seems quite possible that some regions, because the highway is built through them or near them, will actually suffer monetary losses. Generally, this will be true of places or areas already in a subservient position to a larger and more prosperous place. Yet owing partly to the advantages of centralization, the benefit for all areas is positive. Thus, the problem becomes one of possibly restoring the balance through additional investment, perhaps of a different sort, or through a tax adjustment.

TABLE 14-II
ALLOCATION OF BENEFITS: NINE-REGION CASE, 1950

Area	Producers' Gain or Loss*	Consumers' Gain or Loss*	Per Physician†	Per 1,000 Customers‡
East North Central	$-52,000	$-47,120	$-427	$-140
Middle Atlantic	86,020	37,980	450	113
New England	7,120	10,300	131	106
All producers	41,140		112	
All consumers		560		neg.‡
Net social pay-off	41,700			
East South Central	-1,000	-2,430	-37.50	-18.90
South Atlantic	2,018	2,332	26.80	9.70
All producers	1,018		10.00	
All consumers		-336		neg.‡
Net social pay-off	682			
U.S. producers	42,158		66.00	
U.S. consumers		124		neg.‡
NET SOCIAL PAY-OFF	42,282			

*Unadjusted values.
†Adjusted values.
‡Negligible.

Western Pennsylvania Pilot Study

The existence of a large body of data for obtaining an equilibrium solution and for comparison with actual empirical observations makes Western Pennsylvania a valuable area. The solution for Western Pennsylvania (1950) is summarized in Table 14-III and on maps (Figures 14-2, 14-3). These give the results of a before and after analysis, if the construction of the planned links in the Interstate Highway System is assumed.

Interpretation of Basic Equilibrium Solution

The dominance of Pittsburgh is of course expected and is here demonstrated. Note particularly the extension of Pittsburgh's influence eastward along the Penn-

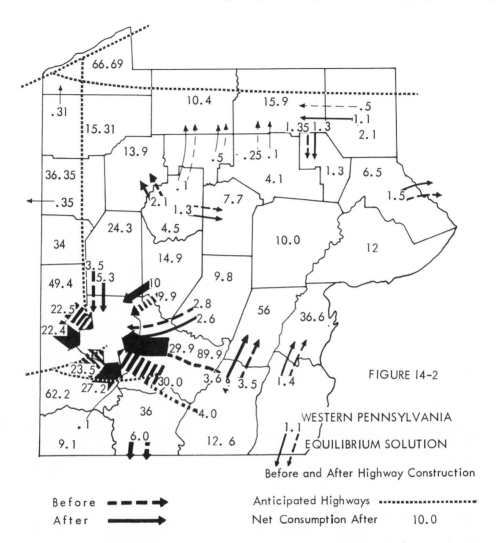

FIGURE 14-2

WESTERN PENNSYLVANIA

EQUILIBRIUM SOLUTION

Before and After Highway Construction

Before ▬ ▬ ▬ ➤   Anticipated Highways ▪▪▪▪▪▪▪▪▪▪▪▪▪▪▪▪▪▪

After   ▬▬▬▬➤   Net Consumption After   10.0

Note: Outward Flows Represent Increased Area Consumption

sylvania turnpike. There are several adjoining counties in neighboring states which supply parts of the area--Williamsport, Pa.; Cumberland, Md.; Morgantown, W. Va.; and Youngstown, Ohio. Other places emerge as regional centers of surplus care--Blair, Cambria, Erie, McKean, Jefferson, and Venango counties. An estimated 60 additional physicians are visited through movement out of the area, as to Buffalo, Philadelphia, New York, and elsewhere.

Interpretation of Changes Following Highway Construction

The hypothetical solution after highway construction is not greatly different from the before solution. Movement is now permitted between Erie and Crawford. The new east-west highway across the northern part of the area permits Potter

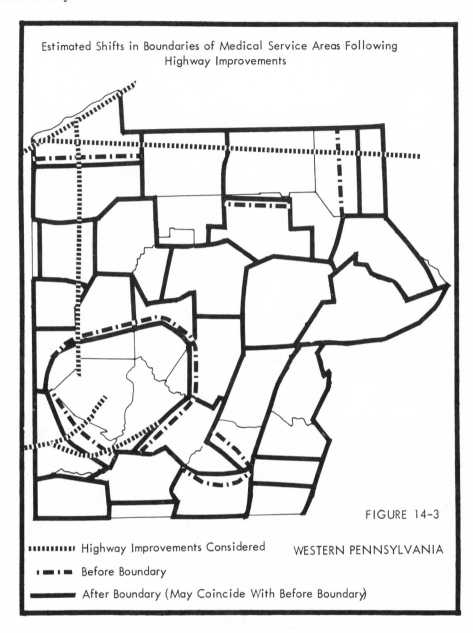

Estimated Shifts in Boundaries of Medical Service Areas Following
Highway Improvements

FIGURE 14-3

••••••••• Highway Improvements Considered   WESTERN PENNSYLVANIA

▪▬▪▬▪▬▪ Before Boundary

▬▬▬▬▬▬ After Boundary (May Coincide With Before Boundary)

an increased share of McKean's surplus at the expense of counties not on the highway. The new highway to be built southwest from Pittsburgh to Wheeling aids Washington County to increase its volume of movement at the expense of other counties adjacent to Pittsburgh, but effectively raises total exports of the city. Total movements for all counties increase from 109.15 to 112.9, in terms of physicians.

Delimitation of Service Areas

In all, 26 of 29 counties are involved in movements, indicating that counties or groups of counties are not realistic medical service areas. In Figure 14-3 are es-

TABLE 14-III
EQUILIBRIUM SOLUTION: WESTERN PENNSYLVANIA, 1950

	Specialists		Internal Price	Equil. Price		Net Consumption		Trade (Physicians)	
	Supply	Demand		Before	After	Before	After	Before	After
Allegheny	670	564.0	-7.0	-0.5	-0.3	572.00	569.0	-98.00	-101.00
Armstrong	5	27.0	27.2	14.8	15.0	15.00	14.9	10.10	9.90
Beaver	27	64.5	21.4	8.5	8.7	49.50	49.4	22.50	22.40
Bedford	2	10.5	20.8	14.6	14.6	4.50	4.5	2.50	2.50
Blair	38	45.0	5.0	6.0	6.0	36.60	36.6	-1.40	-1.40
Butler	19	33.0	14.4	8.8	9.0	24.50	24.3	5.50	5.30
Cambria	60	69.5	4.5	6.2	6.4	56.50	56.0	-3.50	-4.00
Cameron		2.5	35.7	16.3	16.0	1.35	1.3	1.35	1.30
Centre	12	20.5	12.9	12.9	12.9	12.00	12.0		
Clarion	1	12.0	28.8	19.6	19.6	4.50	4.5	3.50	3.50
Clearfield	10	26.0	18.6	18.6	18.6	10.00	10.0		
Clinton	5	12.0	19.2	15.0	15.0	6.50	6.5	1.50	1.50
Crawford	15	26.0	13.9	13.9	13.5	15.00	15.3		0.31
Elk	4	12.0	23.2	22.3	23.0	4.25	4.1	0.25	0.10
Erie	67	81.0	6.4	6.4	6.5	67.00	66.7		-0.31
Fayette	30	60.5	16.1	13.0	13.0	36.00	36.0	6.00	6.00
Forest		1.5	30.6	20.3	20.3	0.50	0.5	0.50	0.50
Greene	9	14.5	12.1	12.1	12.0	9.00	9.1		0.10
Indiana	7	23.5	21.4	17.8	18.0	9.80	9.6	2.80	2.60
Jefferson	9	15.0	12.2	14.9	14.9	7.70	7.8	-1.30	-1.30
Lawrence	34	38.5	4.3	4.3	4.3	34.00	34.0		
McKean	18	20.0	3.5	7.3	8.0	15.75	15.5	-2.25	-2.50
Mercer	36	42.0	5.4	5.0	5.0	36.34	36.4	0.35	0.35
Potter	1	4.5	20.8	17.7	14.4	1.50	2.1	0.50	1.10
Somerset	5	24.5	23.8	14.5	14.7	12.50	12.6	7.50	7.60
Venango	16	21.5	8.4	11.9	11.9	13.90	13.9	-2.10	-2.10
Warren	11	14.0	7.0	8.6	8.6	10.40	10.4	-0.60	-0.60
Washington	35	71.5	17.4	6.1	4.7	58.50	62.2	23.50	27.20
Westmoreland	60	107.5	15.7	5.5	5.7	90.00	89.9	30.00	29.90
Western Pa.	1,206	1,466	6.2	6.2	6.2	1,215	1,215	± 109.15	± 112.90
Adjacent counties								-9	-9
Rest of East*								-60	-60

*Estimated movement out of area not included in above figures.

timated, quite grossly, the service areas, as indicated by the analysis, for a medium-low level of specialization and for before and after highway construction. These should be compared with physicians' service areas as derived from two other approaches and shown in Figure 14-4. That of the American Medical Association is based on a best guess and is more specifically geared to particular cities, while that obtained by Ciocco and Altman (1951) from the same data on which parts of this study are based (but using a percentage index for assigning counties to service areas) is similar but suffers from dependence upon county lines.

The largest area, of course, is served by Pittsburgh, including parts of 9 of the 29 counties. Note especially the elongation of the service area along the turnpike east of Pittsburgh. Other centers serving more than one county are McKean, Blair, Cambria, Erie and Venango and Jefferson, which apparently split Clarion county in so far as this level of specialization is concerned. Pittsburgh is the only really dominant center. The rest of the area, for a number of reasons, is broken into many service areas. Apparently the existence of Pittsburgh discourages the growth of other large service centers, but it is not so easily accessible as to dominate directly the entire area. Finally, note the changes in the extent of service

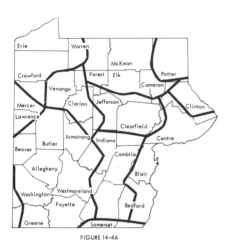

FIGURE 14-4A

Delineation of Medical Service Areas (Dickinson - American Medical Association)

FIGURE 14-4B

Delineation of Physician Service Areas (Ciocco and Altman)

areas after construction of the highways, expanding in the direction of increased trade, and alternately contracting. For example, the construction of the highway through Washington County increases Pittsburgh's penetration there and indirectly increases the influence of Cambria on Somerset by contracting Pittsburgh's influence to the east.

Comparison with Empirical Observations

Figure 13-1 presented the movements of persons to physicians as provided by sample observations in 1950. The results are quite similar to those in the present analysis but with various local and generally minor exceptions. These observations are based on all physicians, rather than specialists as above, and include some local movements, as from Cambria to Blair, which cannot occur in the above analysis owing to the reduction of areas to points for transport cost purposes. Movements identified by the observations but not by the model include Jefferson to Allegheny, Clearfield to Center, Elk to Clearfield, Center to Clinton, and from Warren, McKean, and Potter to New York State. Many of these would be identified if regions finer than counties were used. Movements identified by the model but not by observations include Elk to McKean, Bedford to Blair, Indiana to Allegheny, and Clarion to Jefferson. Much of the difference in these cases is due to the number and kinds of physicians who answered the questionnaire while, of course, the model was based on all specialists. Nevertheless, the two patterns are close, even as to proportional magnitude of movements in the case of the counties adjoining Pittsburgh.

Measurement of Benefits

A rather limited amount of Interstate Highway construction is anticipated for Western Pennsylvania, at least in so far as highways will be improved, since the turnpike already exists. Consequently, the benefits and losses are not large, as is evidenced in Table 14-IV. Of course, the analysis provides breakdowns for every county as well as the summary areas of the table.

TABLE 14-IV
ALLOCATION OF BENEFITS

Gain or Loss to	Allegheny Complex	Erie- Crawford	Potter- McKean	Total	Per Physician*	Per 1,000†
Exporter producer	$214.95	$9.40	$11.50	$235.85	$20.75	
consumer	200.25	25.80	19.15	245.20		$6.70
area net	415.20	35.20	30.65	481.05		
Importer producer	10.05	-36.60	-1.00	-26.55		
consumer	-178.60	34.50	-27.05	-171.15		-4.65
area net	-168.55	-2.10	-28.05	-197.70		
All producers	225.00	-26.20	10.50	209.30	18.35	
All consumers	21.65	60.15	-7.90	73.90		2.05
Net social pay-off	246.65	34.20	2.60	283.20		

*Per physician of affected areas.

†Per 1,000 population of affected areas.

Important note: The first four columns of figures are direct and unadjusted results, and the final two adjusted figures, as in Table 14-II.

Although not large in magnitude, the patterns of benefits and losses or indirect costs are highly interesting. The following conclusions can be made from the analysis:

1. For a service such as physician care, in which patients move to physicians and absorb the cost of transportation, an improvement in access generally leads to greater producer than consumer monetary benefits. These are largely nonvehicular benefits.

2. Exporting or surplus regions have benefits to both producers and consumers, while importing or deficit areas often have greatly increased consumer expenditures (relative economic loss) and generally, but not necessarily, producer losses as well. These are interpretable as reorganization or nonvehicular benefits.

2A. In a social sense, the interpretation may differ. The surplus region consumer, given a higher price, consumes less care; the deficit region consumer, given a lower price, consumes more, which results in improved care. However, movement toward an equilibrium *always* evens out the per-family level of medical care among regions and often but not necessarily results in a net total saving in expenditures, even after transportation cost is charged to consumers.

3. The improvement in highways involves a larger flow of services often at a lower absolute cost, always at a lower unit cost.

4. The improvement in a highway affects all areas or counties that are interrelated in trade rather than just those through which the highway passes.

5. An improvement to one link (export-import) of a set of areas or counties increases total trade significantly in the import (deficit) area and tends to raise all prices except in that area. For the deficit area the effect is to strengthen the economic tie, which discourages local producer growth and results in a net increased flow of income to the surplus center. In reference to urbanized

areas, such an improvement will help the central cities recover income lost by the suburbanization of wealthy residents, but outside of urban areas, the effect is to aggravate the economic differences between surplus-exporting towns and cities, and smaller importing towns and rural areas.

Seattle Equilibrium of Physician Utilization

The Seattle area likewise has been chosen for an urban analysis since data for completion of an equilibrium solution are available and some comparative empirical observations as well. The region has been subdivided into a large number of areas which are census tracts or groups of census tracts. Since there are 102 such areas, the mathematical task is considerably increased. Other special factors are the extreme dominance of the downtown-First Hill areas, so that an extreme outlying area is often served both from downtown and from an outlying business center. Further, the proximity yet distinctness of downtown and First Hill complicates the problem.

A full table of results would be extremely long. Consequently, the results will be summarized only on maps (Figures 14-5, 14-6). It is necessary to omit most information, presenting only the equilibrium consumption in terms of physicians. These maps summarize the before and after freeway construction respectively and should be examined along with Figures 13-3 and 13-5 which indicate the relative supply and demand of physicians by areas.

Interpretation of Basic Equilibrium Solution
The basic solution consists essentially of a split of the Seattle area into two parts--one served by downtown, the other by First Hill--except for five outlying service areas around Ballard, the University, Northgate, Kirkland-Bellevue, and Capitol Hill. The solution is similar to the preliminary hierarchy portrayed in Figure 13-4, although it is not strictly comparable.

In general, the downtown area serves the western half of the region traversed by U.S. 99, and First Hill the eastern portion. This property of the solution is remarked on again in the list of deficiencies of the model. It is interesting that no area to the south with contiguous land connection has any exports while to the north and east are several separated by former water barriers. A further feature of the solution is the necessity for most areas near the city center to consume an amount of service greater than their demand. The close-in saturation is due to the effect of higher prices in outlying areas which results in consumption lower than demand, while for the entire area the surplus of physicians is barely relieved by exports out of the area (Table 14-V).

There are two large service areas, then, interrupted by a few enclosed outlying service areas. In accordance with theory, these areas extend outward only from the surplus center (for example, Capitol Hill). The pattern is clearly not an optimal equilibrium according to theory, however, since overlap of market areas occurs. Part of the answer lies in the use of all physicians in the calculations, in which case we might interpret the overlapped areas as service areas for some lower level of specialization in contrast to the entire area as a higher level specialist one. Nevertheless, the empirical observations show that movement occurs,

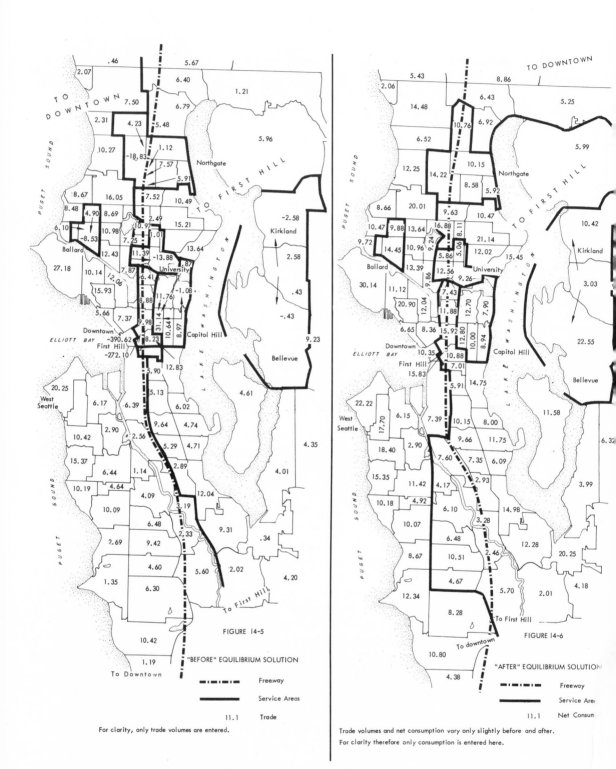

FIGURE 14-5

"BEFORE" EQUILIBRIUM SOLUTION

	Freeway
	Service Areas
11.1	Trade

For clarity, only trade volumes are entered.

FIGURE 14-6

"AFTER" EQUILIBRIUM SOLUTION

	Freeway
	Service Area
11.1	Net Consu

Trade volumes and net consumption vary only slightly before and after.
For clarity therefore only consumption is entered here.

TABLE 14-V
SUMMARY: SEATTLE EQUILIBRIUM SOLUTION

Area	Physician Supply	Physician Demand	Net Consumption		Trade	
			Before	After	Before	After
Importing areas	217	889.85	887.97	888.98	670.97	671.98
Exporting areas	869	126.60	129.89	128.81	-739.71	-740.19
Downtown	401	24.05	26.59	26.62	-662.01	-662.38
First Hill	288					
Capitol Hill	53	19.85	20.78	20.62	-32.22	-32.38
Outlying	127	79.10	81.52	81.57	-45.48	-45.43
Seattle area	1,086	1,016.45	1,017.26	1,017.79		
Outside area					68.74*	68.21

*Estimated 6 per cent use of physicians by out of area residents.

by necessity, to the downtown from beyond competing centers which indicates an underlying nonoptimal equilibrium.

Deficiencies

It is necessary, however, to recognize major deficiencies in the use of this model unmodified, for the Seattle case.

1. The unreality of the normal equilibrium sustained by *price* in an urban area. In accordance with theory, differences in prices at equilibrium in the solution depend on the structure of transport costs. Obviously, the lowest prices occur downtown and on First Hill. This situation is responsible for the inability of an outlying center to export in a central direction. In reality, however, central movement does occur (Figure 13-7) and there is an almost constant producers' price over the entire area. Such a restriction could be built into the model although the solution would not be an optimal equilibrium (the downtown area absorbing the costs of nonoptimality). The effect in this case would be to reshape the outlying service areas to include the closest areas in any direction. This would relieve the unreality of never permitting service areas to extend inward from an outlying surplus point. On the other hand, the gateway principle is still valid and often recognized empirically, which indicates that perhaps price differences (even though hidden) are real. At this time it is not clear what modifications could be made from the two extreme positions.

2. The inability of persons to distinguish small cost differences. The empirical observations indicate, above all, that the service areas as identified for downtown and First Hill are unjustified. More people on the west side tend to use the downtown and vice versa, but clearly the difference in distance of a little over one-half mile is not sufficient to discourage significant crossing-over from one to the other. In the solution the ability to control the market of some outlying center often hinged on a very small amount, as .3 miles, and obviously in such cases the consumer treated the downtown-First Hill areas as one.

A further problem in this connection is the area served by surplus areas on Capitol Hill. One is the result of the Group Health Clinic, which is a medical subscription plan center drawing on members not at random or locally but primarily from all over the south side. The other is a specialized psychiatric clinic. These, of course, could have been removed from the analysis altogether. The inability to distinguish small price differentials also plays a part in the decision of whether to move to the downtown area or to a local center. In reality some persons would probably prefer to go a shorter distance with less traffic than downtown, even if the local price were higher. This is borne out strongly by the distribution of sample visits to Northgate physicians (Figure 13-6). How may we incorporate into a model this lack of price discrimination? One way would be to add the restriction that movement is to the nearest surplus center so long as the price was not, say, 5 per cent greater than that quoted from a more distant surplus center. Another would be to permit a range of indecision, for example, 2 per cent of each side of equilibrium prices as interpreted by customers. This would cause, likewise, a range of solutions which might merely change a boundary line into a zone of indifference or perhaps make a whole area indifferent as to which producer it would utilize. Indifference suggests, in fact, utilization of both, which is more realistic according to the sample observations reported earlier.

3. The limitation of given regions. The use of indivisible regions, generalized to points, forces an unreal configuration of service areas (for example, around Ballard). There is available a solution to a continuous spatial model, but it is operationally too difficult or impossible for a problem such as the present one (Beckmann, 1952). However, if we assume that a small area is continuously deficit throughout, it would be possible in the present analysis to permit a divisibility of regions by permitting a continuous variability of transport costs depending on the penetration of an area. This increases the work mathematically but not to the extent of Beckmann's solution, in which production and consumption vary continuously over space as well.

Interpretation of Changes Incident to Freeway Construction

The construction of the freeway, under the limitations of this model noted above, has interesting consequences. Time-cost distances are shortened greatly, but especially to First Hill, for areas on or close to the freeway. This has the immediate effect of making First Hill more competitive in service areas formerly served from downtown. If producers could move locations, which of course occurs in reality, we would expect a gradual movement of physicians to First Hill from downtown, adjusting in this way to the changed competitive situation. This can be handled within the model, but in the present example, since production location is given and does not change, the downtown area must lower its prices in order to balance supply and demand. In this way it, in turn, becomes more competitive in some other areas in the north. These changes must be considered only as a tendency in accordance with the limitations of the model noted above. The only other area change is a substitution of areas in the service area of Northgate. As expected, trade increases in all areas directly served by the freeway and declines in most others, except for a few near the freeway. Total trade increases but slightly in this example, due to saturation of demand.

Measurement of Benefits

The benefits and losses or indirect costs as a result of hypothetical freeway con-

TABLE 14-VI
SEATTLE: SUMMARY OF DOLLAR BENEFITS

Item	Amount Unadjusted	Adjusted Benefit	
		Per Physician	Per 1,000 Population
Transport cost savings	$427.07		$55.00
Exporter's gain (trade)	65.77	$6.64	
Outlying centers	-22.73		
Capitol Hill	25.06		
Downtown	-300.97		
First Hill	424.42		
Net export areas benefit	65.77		
Consumer loss (trade)	-76.28		-22.92
Net consumer area benefit	250.79		32.65
Net social pay-off	316.56		

struction and no changes in service location are summarized in Table 14-VI. In contrast to the previous examples and because of the pre-existent saturation of demand, trade did not increase much, and consequently exporting areas' (producers') benefits were small. The gain, therefore, was realized mostly as consumer transport cost savings (vehicular benefits) which much more than offset normal consumer trade losses in importing areas.

The benefits and losses of specific areas are not listed owing to the extreme length of such a table. Very few areas had increased transport costs. Areas with less trade had automatic declines in cost, and those with increased trade had a correspondingly greater cut in unit transport cost. In agreement with theoretical remarks made in the introduction, outlying producing areas all had net losses while First Hill and Capitol Hill gained. In this way freeway construction aids centralization of business. These gains and losses seem extremely small (for example, less than 4 cents per capita per year), but if it is realized that only one-fifth of 1 per cent of consumer expenditures is for physician care, we can see that if other industries behaved similarly, benefits would be in the neighborhood of $20.00 per capita per year. It has already been noted that a significant margin of error is associated with these estimates. On the other hand, the inelasticity of demand for medical care suggests that the estimates may be conservative when generalized to all industries.

General Conclusion

This study has undertaken a triple task of measurement: (1) Changes in production, consumption, and trade between areas resulting from highway construction, (2) changes in service areas, and (3) benefits, including gains or losses (indirect costs) to areas and to producers and consumers. The case of physicians' services was used because of the availability of data, but information problems existed even here. The analysis indicated that as between larger areas the method is quite satis-factory, although in this paper changes in service areas were not treated with pre-cision. Apparently, urban areas present certain problems which render the present analysis less useful, but the discussions hold hope for successful modification in this area.

Probably the most interesting part of the analysis, in terms of highway studies, is the demonstration of a method for measuring benefits. Also, general statements may be made from the analysis.

The summary statements are:

1. Producers generally gain most from highway construction where trade and con-sumption may increase. Consumers pay more, but receive more care and pos-sibly better care. These are nonvehicular benefits.

2. Where trade and consumption are relatively inelastic, consumers gain most through transport cost savings. These are primarily vehicular benefits.

3. New highways provide benefits and easier access to central cities, or major nodes, at the expense of outlying centers and rural territory. This helps cen-tral cities to retain expenditures of wealthy suburban residents but serves to aggravate the relative disparity between rural-small town and large urban cen-ter income at a regional and national scale. At the same time it is possible for rural-small town areas to consume more easily the more specialized ser-vices available in larger cities. Higher levels of production, use of transporta-tion, and consumption accompany transportation developments. The point here is that locations and groups show these changes unequally. Equally important is the fact that reorganization and nonvehicular benefits and associated costs exist and can be allocated to areas or groups.

Bibliography

Alderson and Sessions (1957). "Basic Research Report on Consumer Behavior." Philadelphia: Alderson and Sessions. (Mimeographed.)

American Dental Association (1956). "Survey of Patient to Dentist Travel," American Dental Association, *Journal*, 53:461-66.

American Medical Association (1951 & 1957). *American Medical Directory--A Register of Physicians*. 18th and 19th Editions. Chicago: American Medical Association.

Applebaum, William (1932). *The Secondary Commercial Centers of Cincinnati*. Cincinnati: Institue of Industrial Research, University of Cincinnati.

------- (1956). *A Quarter-Century of Change in Cincinnati Business Centers*. Cincinnati: The Cincinnati Enquirer.

-------, and B. L. Schapker (1955). *Atlas of Business Centers: Cincinnati, Hamilton County*. Cincinnati: The American Marketing Association.

Arisue, Takeo (1957). "A Regional Study on Passenger Travel in Japan," *Geographical Review of Japan*, 30:1016-29.

Aubert-Krier, J. (1949). "Monopolistic and Imperfect Competition in Retail Trade," *Monopoly and Competition and Their Regulation*, ed. E. Chamberlin. London: Hutchinson's University Library.

Bachman, George W. (1952). *Health Resources in the United States*. Washington: Brookings Institution.

Baker, G., and B. Funaro (1951). *Shopping Centers*. New York: Reinhold and Co.

Barkley, Robert E. (1951). Origin-Destination Surveys and Traffic Volume Studies. Washington: Highway Research Board, *Bibliography No. 11*.

Bartholomew, Harland (1955). *Land Uses in American Cities*. Cambridge: Harvard University Press.

-------, and Associates (1950a). "Parking Facilities in the Central Business District, Cedar Rapids, Iowa," St. Louis: Harland Bartholomew and Associates. (Mimeographed.)

-------, and Associates (1950b). *A System of Major Streets, Cedar Rapids, Iowa*. St. Louis: Harland Bartholomew and Associates.

-------, and Associates (1953). *Land Use and Amended Zoning Ordinance, Cedar Rapids, Iowa*. St. Louis: Harland Bartholomew and Associates.

-------, and Associates (1957). *Parking Facilities in the Central Business District*. St. Louis: Harland Bartholomew and Associates.

277

Baumol, W. J., and E. A. Ide (1956). "Variety in Retailing, " *Management Science*, 3:93-101.

Beckmann, Martin J. (1952). "A Continuous Model of Transportation, " *Econometrica*, 20:643-66.

------- (1953). "The Partial Equilibrium of a Continuous Space Market, " *Weltwirtschaftliches Archiv*, 71:Heft 1, 73-87.

------- (1958). "City Hierarchies and the Distribution of City Size, " *Economic Development and Cultural Change*, 6:243-48.

-------, and T. Marschak (1956). *An Activity Analysis Approach to Location Theory*. (Cowles Foundation Paper No. 99.) New Haven: Cowles Foundation for Research in Economics.

-------, C. B. McGuire, and Christopher B. Winsten (1956). *Studies in the Economics of Transportation*. New Haven: Yale University Press.

Belcher, John C. (1955). "The Changing Distribution of Medical Doctors in Oklahoma. " Oklahoma A & M. College, Agricultural Experiment Station, *Bulletin B-459*.

------- (1956). "Medical Service Area Relationships in Harper County, Oklahoma, " Oklahoma A & M College, Agricultural Experiment Station, *Bulletin B-477*.

Berry, Brian J. L. (1956). "Geographic Aspects of the Size and Arrangement of Urban Centers." M.A. Thesis, University of Washington.

------- (1958). "Shopping Centers and the Geography of Urban Areas. " Ph. D. Dissertation, University of Washington.

-------, and William L. Garrison (1958a). "Functional Bases of the Central Place Hierarchy, " *Economic Geography*, 34:145-54.

-------, and William L. Garrison (1958b). "Recent Developments of Central Place Theory, " Regional Science Association, *Papers and Proceedings*, 4:107-20.

-------, and William L. Garrison (1958c). "A Note on Central Place Theory and the Range of a Good, " *Economic Geography*, 34:304-11.

Blumenfeld, H. (1949). "On the Concentric Circle Theory of Urban Growth, " *Land Economics*, 25:208-12.

Bostick, Thurley A., Roy T. Messer, and Clarence A. Steele (1954). "Motor-Vehicle-Use Studies in Six States, " *Public Roads*, 28:99-125.

Breese, Gerald William (1949). *The Daytime Population of the Central Business District of Chicago*. Chicago: University of Chicago Press.

Brewster, Maurice R., William A. Flinn, and Ernest H. Jurkat (1955). *How to Make and Interpret Locational Studies of the Housing Market*. Washington: Government Printing Office.

Brink, E. L., and J. S. de Cani (1957). "An Analogue Solution of the Generalized Transportation Problem With Specific Application to Marketing Location, " First International Conference on Operational Research, *Proceedings*. Baltimore: Operations Research Society of America.

Brownlee, O. H., and Walter W. Heller (1956). "Highway Development and Financing, " *American Economic Review*, 46:232-50.

Burch, James S. (1957). "The Secondary Road Program in North Carolina, " Highway Research Board, *Bulletin 147*.

Burgess, E. W. (1923). "The Growth of the City, " American Sociological Society, *Proceedings*, 18:85-89.

Burros, R. H. (1957). "A Mathematical Theory of Static Hierarchy." Paper read before Operations Research Society of America, Annual Meeting, May 1957.

Canoyer, H. G. (1946). *Selecting a Store Location.* (U.S. Dept. of Commerce, Economic Series No. 56.) Washington: Government Printing Office.

Carroll, J. Douglas. Jr., and Howard W. Bevis (1957). "Predicting Local Travel in Urban Regions," Regional Science Association, *Papers and Proceedings*, 3: 183-97.

Cassady, R., and H. Ostlund (1935). *Retail Distribution Structure of the Small City.* (Studies in Economics and Business, No. 12.) Minneapolis: University of Minnesota Press.

Chamberlin, E. H. (1933). *Theory of Monopolistic Competition.* Cambridge: Harvard University Press.

------- (1957). *Towards a More General Theory of Value.* New York: Oxford University Press.

Chittick, D. (1955). "Growth and Decline of South Dakota Trade Centers, 1901-1951," South Dakota State College, Rural Sociology Dept., Agricultural Experiment Station, *Bulletin 448.*

Christaller, W. (1933). *Die zentralen Orte in Süddeutschland.* Jena: Gustav Fischer Verlag.

Ciocco, Antonio, and Isidore Altman (1954). "Medical Service Areas as Indicated by Intercounty Movements of Patients," and "Distances Traveled for Physicians in Western Pennsylvania," Parts 1 and 2, U.S. Public Health Service, *Monograph 19.* Washington: Government Printing Office.

Connor, Ruth M., and William G. Mather (1948). "Use of Health Services in Two Southern Pennsylvania Communities," Pennsylvania State College, Agricultural Experiment Station, *Bulletin 504.*

Converse, P. D. (1949). "New Laws of Retail Gravitation," *Journal of Marketing,* 14:329-85.

Copeland, M. T. (1922-23). "Relation of Consumers' Buying Habits to Marketing Methods," *Harvard Business Review,* 1:282-89.

Courtney, Robert (1955). "Regional Reporter," April, 1955, cited in Supplement No. 1, *Ontario Planning,* Vol. 2, No. 7.

Curtiss, C. D. (1957). "Urban Highway Planning: Its Increasing Importance," *Traffic Quarterly,* 2:445-57.

Davenport, D. H. (1927). "The Retail Shopping and Financial Districts of New York and Its Environs." *Regional Survey of New York and Its Environs.* New York: Regional Plan of New York and Its Environs.

Dearing, Charles L. (1957). "Toll Roads, Rates and Highway Pricing," *American Economic Review,* 47:441-52.

De Geer, Sten (1923). "Stockholm: A Geographical Appraisal," *Geographic Review,* 13:487-500.

Dessel, M. D. (1957). *Central Business Districts and Their Metropolitan Areas.* (U.S. Dept. of Commerce, Area Trend Series No. 1.) Washington: Government Printing Office.

Dickinson, Frank (1950). "A Comparison of State Physician-Population Ratios for 1938 and 1949," American Medical Association, *Bulletin 78.*

------- (1951). "Supply of Physicians' Services," American Medical Association, *Bulletin 81.*

------- (1954). "Distribution of Physicians by Medical Service Areas," American Medical Association, *Bulletin 94.*

------- (1956). "A Study of Medical School Alumni," American Medical Association, *Journal,* 160:473-84.

-------, and Charles E. Bradley (1951). "Medical Service Areas," American Medical Association, *Bulletin 80.*

Dickinson, Robert E. (1947). *City, Region, and Regionalism.* London: Kegan Paul, Trench, and Trubner.

------- (1957). "The Geography of Commuting: The Netherlands and Belgium," *Geographical Review,* 47:521-38.

Dorau, H. B., and A. E. Hinman (1928). *Urban Land Economics.* New York: Macmillan.

Duesenberry, James S. (1949). *Income, Saving and the Theory of Consumer Behavior.* Cambridge: Harvard University Press.

Duncan, Beverly (1957). "Intra-Urban Population Movement," *Cities and Society,* ed. P. K. Hatt and A. J. Reiss, Jr. Glencoe: The Free Press.

Dunn, E. S. (1955). *The Location of Agricultural Production.* Gainesville: University of Florida Press.

Durden, C. Dennis (1955). "Some Geographic Aspects of Motor Vehicle Travel in Rural Areas--Empirical Tests of Certain Geographic Concepts of Location and Interaction." Ph.D. Dissertation, University of Washington.

Ely, R. S., and G. S. Wehrwein (1940). *Land Economics.* New York: Macmillan.

Enke, S. (1951). "Equilibrium Among Spatially Separated Markets: Solution by Electric Analogue," *Econometrica,* 19:40-47.

Fein, Rashi (1954). "Studies of Physician Supply and Distribution," *American Journal of Public Health,* 44:611-24.

Ferber, Robert (1955). "Expenditures for Services in the United States: Their Growth and Some Factors Influencing Them." Paper read before the Conference on Consumption and Economic Development of the Universities--National Bureau Committee for Economic Research.

Ferguson, Allen R. (1958). "A Marginal Cost Function for Highway Construction," *American Economic Review,* 48:223-34.

Firey, W. (1947). *Land Use in Central Boston.* Cambridge: Harvard University Press.

First National Bank of Phoenix (1957). *Shopping Centers in Greater Phoenix.* Phoenix: First National Bank.

Fisher, Walter D. (1957). "Economic Aggregation as a Minimum Distance Problem," Abst. in *Econometrica,* 25:363.

Foley, D. L. (1957). "The Use of Local Facilities in the Metropolis," *Cities and Society,* ed. P. K. Hatt and A. J. Reiss, Jr. Glencoe: The Free Press.

Fox, Karl A., and Richard Taeuber (1955). "Spatial Equilibrium Models of the Livestock Feed Economy," *American Economic Review,* 45:584-608.

Friedmann, J. R. P. (1955). *The Spatial Structure of Economic Development in the Tennessee Valley.* (Research Paper No. 39.) Chicago: University of Chicago, Department of Geography.

Gallion, A. B. (1950). *The Urban Pattern*. New York: D. Van Nostrand & Co.

Galpin, C. J. (1915). "Social Anatomy of an Agricultural Community," University of Wisconsin, Agricultural Experiment Station, *Research Bulletin 34*.

Gardner, Joseph C., Jr. (1949). "A Study of Neighborhood Travel Habits in Baltimore, Maryland," M.A. Thesis, Cornell University.

Garrison, William L. (1950). "The Business Structure of the Consumer Tributary Area of the Fountain Square Major Outlying Business Center of Evanston, Illinois." Ph.D. Dissertation, Northwestern University.

------- (1954). "Community Business Centers." Unpublished Manuscript, University of Washington, Department of Geography. (Dittoed.)

------- (1956). *The Benefits of Rural Roads to Rural Property*. Seattle: Washington State Council for Highway Research.

------- (1957). "Verification of a Location Model," *Northwestern University Studies in Geography*, 2:133-40.

-------, and Brian J. L. Berry (1957). "A Source of Theory for Highway Impact Studies," *Economic Impact of Highway Improvement*. Washington: Highway Research Board, *Special Report 28*.

-------, and Duane F. Marble (1957). "The Spatial Structure of Agricultural Activities," Association of American Geographers, *Annals*, 47:137-44.

-------, and Duane F. Marble (1958). "The Analysis of Highway Networks: A Linear Programming Formulation." Washington: Highway Research Board, *Proceedings*, 37:1-14.

-------, and Marion E. Marts (1958a). *Influence of Highway Improvements on Urban Land: A Graphic Summary*. Seattle: University of Washington, Department of Geography and Department of Civil Engineering.

-------, and Marion E. Marts (1958b). *Geographic Impact of Highway Improvement*. Seattle: University of Washington, Department of Geography and Department of Civil Engineering.

Godlund, Sven (1956a). *Bus Service in Sweden*. (Lund Studies in Geography, Series B: Human Geography, No. 17.) Lund: The Royal University of Lund, Department of Geography.

------- (1956b). *The Function and Growth of Bus Traffic Within the Sphere of Urban Influence*. (Lund Studies in Geography, Series B: Human Geography, No. 18.) Lund: The Royal University of Lund, Department of Geography.

Grebler, Leo, David M. Blank, and Louis Winnick (1956). *Capital Formation in Residential Real Estate: Trends and Prospects*. Princeton: Princeton University Press.

Gruen, V., and L. P. Smith (1952). "Shopping Centers," *Progressive Architecture*, 33:67-109.

Haig, R. M. (1926). "Toward an Understanding of the Metropolis," *Quarterly Journal of Economics*, 40:179-208, 402-34.

------- (1927). "Major Economic Factors in Metropolitan Growth and Arrangement." *Regional Survey of New York and Its Environs*. New York: Regional Plan of New York and Its Environs.

Hall, Edward M. (1958). "Travel Characteristics of Two San Diego Subdivision Developments." Paper presented at the 37th Annual Meeting of the Highway Research Board, Washington.

Hamburg, John R. (1957). "Some Social and Economic Factors Related to Intra-City Movement." M. A. Thesis, Wayne State University.

Harris, Chauncy D. (1943). "Suburbs," *American Journal of Sociology*, 49:1-13.

-------, and Edward L. Ullman (1945). "The Nature of Cities," American Academy of Political and Social Science, *Annals*, 242:7-17.

Hassinger, Edward, and Robert McNamara (1956). "The Pattern of Medical Services for Incorporated Places of 500 or More Population in Missouri, 1950," *Rural Sociology*, 21:175-78.

Hawley, A. H. (1950). *Human Ecology*. New York: Ronald Press.

Heady, E. O. (1952). *Economics of Agricultural Production and Resource Use*. New York: Prentice-Hall.

Hennes, R. G. (1956). *Sharing the Costs of Roads and Streets in Washington*. Seattle, Washington State Council for Highway Research.

Hoffer, Charles R. (1928). "A Study of Town-Country Relationships," Michigan State College, Agricultural Experiment Station, *Special Bulletin 181*.

-------, *et al.* (1950). "Health Needs and Health Care in Michigan," Michigan State College, Agricultural Experiment Station, *Special Bulletin 365*.

Holton, R. H. (1957). "Price Discrimination at Retail: The Supermarket Case," *Journal of Industrial Economics*, 6:13-32.

Hoover, E. (1948). *The Location of Economic Activity*. New York: McGraw-Hill.

Hoyt, H. (1933). *One Hundred Years of Land Values in Chicago*. Chicago: University of Chicago Press.

------- (1939). *Structure and Growth of Residential Neighborhoods in American Cities*. Washington: Government Printing Office.

------- (1949). "Market Analysis of Shopping Centers," Urban Land Institute, *Technical Bulletin No. 12*.

Hunter, Holland (1957). *Soviet Transportation Policy*. Cambridge: Harvard University Press.

Hurd, R. M. (1911). *Principles of City Land Values*. 3rd ed. New York: The Record and Guide.

Iklé, Fred. C. (1954). "Sociological Relationship of Traffic to Population and Distance," *Traffic Quarterly*, 8:123-36.

Iowa State Highway Commission (1958). *Cedar Rapids-Marion Urban Area Origin and Destination Traffic Study*. (Highway Planning Survey, Safety and Traffic Department.) Ames: Iowa State Highway Commission.

Isard, Walter (1956). *Location and Space Economy*. New York: John Wiley & Sons.

-------, and V. Whitney (1949). "Metropolitan Site Selection," *Social Forces*, 27:263-69.

Johnson, E. S. (1957). "The Functions of the Central Business District in the Metropolitan Community," *Cities and Society*, ed. P. K. Hatt and A. J. Reiss, Jr. Glencoe: The Free Press.

Jonassen, C. T. (1955). *The Shopping Center Versus Downtown*. Columbus: Ohio State University, Bureau of Business Research.

Kalaba, R. E., and M. L. Juncosa (1956). "Optimal Design and Utilization of Communication Networks," *Management Science*, 3:33-44.

Kant, Edgar (1957). "Suburbanization, Urban Sprawl and Commutation: Examples from Sweden," *Migration in Sweden: A Symposium*, ed. D. Hannerberg, T. Hagerstrand and B. Odeving. (Lund Studies in Geography, Series B: Human Geography, No. 13.) Lund: The Royal University of Lund, Department of Geography.

Kantner, John (1948). "The Relationship Between Accessibility and Socio-Economic Status of Residential Lands, Flint, Michigan," University of Michigan, Horace H. Rackham School of Graduate Studies. (Mimeographed.)

Kelley, E. J. (1955). "Retail Structure of Urban Economy," *Traffic Quarterly*, 9:411-30.

------- (1956). *Shopping Centers*. Saugatuck: Eno Foundation for Highway Traffic Control.

Klein, Lawrence R., *et al.* (1954). *Contributions of Survey Methods to Economics*. New York: Columbia University Press.

Kolb, J. H., and R. A. Polson (1933). "Trends in Town-Country Relations," University of Wisconsin, Agricultural Experiment Station, *Research Bulletin No. 117.*

Larson, Olaf (1952). "Differential Use of Health Resources by Rural People," *New York Journal of Medicine,* 52:43-49.

Lefeber, Louis (1958). *Allocation in Space*. Amsterdam: North-Holland Publishing Co.

Leibenstein, H. (1950). "Bandwagon, Snob, and Veblen Effects in the Theory of Consumer's Demand," *Quarterly Journal of Economics,* 64:183-207.

Leisch, J. E. (1958). "Spacing and Location of Interchanges on Freeways in Urban and Suburban Areas." Paper read before Portland Convention, American Society of Civil Engineers.

Lewis, W. A. (1945). "Competition in Retail Trade," *Economica,* 12:202-34.

Liepmann, Kate K. (1944). *The Journey to Work*. New York: Oxford University Press.

Lillibridge, R. M. (1948). "Shopping Centers in Urban Redevelopment," *Land Economics,* 24:137-60.

Lösch, August (1944). *Die räumliche Ordnung der Wirtschaft.* 2nd ed. Jena: Gustav Fischer Verlag. Translated by W. H. Woglom and W. F. Stolper (1954), as *The Economics of Location.* New Haven: Yale University Press.

Lung, V. L. (1955). "A Method for Determining Land Needed for Neighborhood Shopping Centers with Special Reference to Eight Seattle Centers." M.A. Thesis, University of Washington.

McKean, Roland N. (1958). *Efficiency in Government Through Systems Analysis.* New York: John Wiley & Sons.

McKeever, J. R. (1953). "Shopping Centers," Urban Land Institute, *Technical Bulletin No. 20.*

------- (1957). "Shopping Centers Restudied," Urban Land Institute, *Technical Bulletin No. 30.*

McKenzie, R. D. (1953). *The Metropolitan Community.* New York: McGraw-Hill.

McQuitty, L. L. (1957). Elementary Linkage Analysis for Isolating Orthogonal and Oblique Types and Typal Relevancies," *Educational and Psychological Measurement,* 17:207-29.

Marble, Duane F. (1958). "Some Geographic and Economic Consequences of Highway Improvement," 9th Annual Road Builders Clinic, *Proceedings.* Pullman: Technical Extension Services.

Martin, Walter T. (1953). *The Rural-Urban Fringe: A Study of Adjustment to Residence Location.* Eugene: University of Oregon Press.

Mayer, H. M. (1942). "Patterns and Recent Trends of Chicago's Outlying Business Centers, "*Journal of Land and Public Utility Economics,* 18:4-16.

Merry, P. R. (1955). "An Inquiry into the Nature and Function of a String Retail Development. A Case Study of East Colfax Avenue, Denver, Colorado. " Ph. D. Dissertation, Northwestern University.

Mertz, William L. (1957). "A Study of Traffic Characteristics in Suburban Residential Areas, " *Public Roads,* 29:208-12.

-------, and Lamelle B. Hamner (1957). "A Study of Factors Related to Urban Travel, " *Public Roads,* 29:170-74.

Mills and Rockleys, Ltd. (1955). *The Size and Nature of the Poster Audience: (Study II).* Coventry: Mills and Rockleys, Ltd.

Mitchell, R. B. , and C. Rapkin (1954). *Urban Traffic: A Function of Land Use.* New York: Columbia University Press.

Mitchell, R. V. (1939). "Trends in Rural Retailing in Illinois, 1926 to 1938, " University of Illinois, *Bulletin,* Vol. 36, No. 100; also Bureau of Business Research, *Bulletin 59.*

Moore, E. Howard, and Raleigh Barlowe (1955). "Effects of Suburbanization Upon Rural Land Use, " Michigan State University; Agricultural Experiment Station, *Technical Bulletin 253.*

Morrill, Richard L. (1957). "An Experimental Study of Trade in Wheat and Flour Milling Industry." M. A. Thesis, University of Washington.

Mott, Frederick D. , and Milton I. Roemer (1948). Rural Health and Medical Care. New York: McGraw-Hill.

Mott, S. H. , and M. S. Wehrly (1949). "Shopping Centers. An Analysis, " Urban Land Institute, *Technical Bulletin No. 11.*

Mountin, Joseph W. (1942 a, b, c, and 1945). "Location and Movement of Physicians, 1923-1938, " *Public Health Reports,* in 4 parts: (1942a) 57:1363-75; (1942b) 57:1752-61; (1942c) 57:1945-53; (1945) 60:173-85.

-------, and Clifford H. Greve (1950). "Public Health Areas and Hospital Facilities, " U.S. Public Health Service, *Publication 42.*

Nelson, L. (1955). "Rural-Urban Distribution of Hospital Facilities and Physicians, " University of Minnesota, Agricultural Experiment Station, *Bulletin 432.*

Nelson, R. L. (1954). "Conservation and Rehabilitation of Major Shopping Districts, " Washington: Urban Land Institute, *Technical Bulletin No. 22.*

New York Academy of Medicine (1947). *Medicine in the Changing Order.* New York: Commonwealth Fund.

Nichols, J. C. (1945). "Mistakes We Have Made in Developing Shopping Centers, " Washington: Urban Land Institute, *Technical Bulletin No. 4.*

O'Brien, Richard E. (1958). *Socio-Economic Forces and Family Pleasure Travel.* Jefferson City: Missouri Division of Resources and Development.

Odoroff, Maurice E. , and L. M. Abbe (1957a). "Use of General Hospitals: Demographic and Ecologic Factors, " *Public Health Reports,* 72:397-404.

-------, and L. M. Abbe (1957b). "Use of General Hospitals:Factors in Outpatient Visits, " *Public Health Reports,* 72:478-83.

Ogburn, W. F. (1946). "Inventions of Local Transportation and the Patterns of Cities, " *Social Forces*, 24:373-79.

Ohio Department of Highways (1957). "Use of Facilities on Limited Access Highways. " Survey made with the cooperation of the Ohio Turnpike Commission, September, 1957. (Mimeographed.)

Oregon State Highway Department (1949). *Portland Metropolitan Area, Traffic Survey, 1946. Origin-Destination Study, 1949.* (Technical Report No. 49-2.) Salem: Oregon State Highway Department.

Orr, Earle W. (1957). "A Synthesis of Theories of Location, of Transport Rates, and of Spatial Price Equilibrium, " Regional Science Association, *Papers and Proceedings*, 3:61-73.

Owen, Wilfred (1956). *The Metropolitan Transportation Problem.* Washington: The Brookings Institution.

Park, R. E. , E. W. Burgess, and R. D. McKenzie (1925). *The City.* Chicago: University of Chicago Press.

Paxton, Edward T. (1955). *What People Want When They Buy a House.* Washington: Government Printing Office.

Pendley, Laurence C. (1956). "Parking and Buying Habits of a Store's Customers. " Washington: Highway Research Board, *Special Report 11C.*

Pennell, M. Y. , and M. E. Altenderfer (1954). *Health Manpower Source Book,* Section 1: "Physicians, " and Section 4: "County Data. " Washington: Government Printing Office.

Pennsylvania, University of (with U. S. Bureau of Labor Statistics) (1954). *Study of Consumer Expenditures, 1950.* Vol. 16: *Medical Care, Personal Services and Transportation.* Philadelphia: University of Pennsylvania, Wharton School of Finance and Commerce.

Polk, R. L. , and Company (1950). *Polk's Cedar Rapids City Directory.* Omaha: R. L. Polk and Company.

Ponsard, C. (1955). *Economie et Espace.* Paris: Sedes.

Potter, H. (1943). "Neighborhood Shopping Centers, " *Architectural Forum*, 79: 76-78.

President's Committee on the Health Needs of the Nation (1952). *Building America's Health.* Vol. 2: *America's Health Status, Needs and Resources;* and Vol. 3: *A Statistical Appendix.* Washington: Government Printing Office.

Proudfoot, M. J. (1936). "The Major Outlying Business Centers of Chicago. " Ph. D. Dissertation, University of Chicago.

------- (1937a). "The Outlying Business Centers of Chicago, "*Journal of Land and Public Utility Economics*, 13:57-70.

------- (1937b). "City Retail Structure, " *Economic Geography*, 13:425-28.

------- (1938). "The Selection of a Business Site, "*Journal of Land and Public Utility Economics*, 14:370-82.

Quinn, James A. (1950). *Human Ecology.* New York: Prentice-Hall.

Rannells, John (1956). *The Core of the City.* New York: Columbia University Press.

Rapkin, Chester (1953). "An Approach to the Study of the Movement of Persons and Goods in Urban Areas. " Ph. D. Dissertation, Columbia University.

-------, Louis Winnick, and David M. Blank (1953). *Housing Market Analysis: A Study of Theory and Methods.* Washington: Government Printing Office.

Ratcliff, R. U. (1935). "An Examination into Some Characteristics of Outlying Retail Nucleations in Detroit, Michigan." Ph.D. Dissertation, University of Michigan.

------- (1939). "The Problem of Retail Site Selection," *Michigan Business Studies*, Vol. 9. University of Michigan, Bureau of Business Research.

------- (1949). *Urban Land Economics*, New York: McGraw-Hill.

------- (1953). "The Madison Central Business Area," *Wisconsin Commerce Papers*, 1: No. 5. University of Wisconsin, Bureau of Business Research and Service.

------- (1954). "The Dynamics of Efficiency in the Locational Distribution of Urban Activities." Columbia University Bicentennial Conference, No. 1. (Mimeographed.)

------- (1955). "Efficiency and the Location of Urban Activities," *The Metropolis in Modern Life*, ed. R. M. Fisher. New York: Doubleday.

------- (1957). "On Wendt's Theory of Land Values," *Land Economics*, 33:360-63.

Rolph, I. K. (1929). *The Locational Structure of Retail Trade*. Vol. 80, Domestic Commerce Series. U.S. Bureau of Foreign and Domestic Commerce. Washington: Government Printing Office.

Rossi, Peter H. (1955). *Why Families Move*. Glencoe: The Free Press.

Saal, Carl C. (1955). "Operating Characteristics of a Passenger Car on Selected Routes." Washington: Highway Research Board, *Bulletin 107*, 1-34.

St. Clair, G. P. (1947). "Suggested Approaches to the Problem of Highway Taxation." Washington: Highway Research Board, *Proceedings*, 1947, 1-6.

Samuelson, Paul A. (1952). "Spatial Price Equilibrium and Linear Programming," *American Economic Review*, 42:283-303.

Schmidt, R. E., and M. E. Campbell (1956). *Highway Traffic Estimation*. Saugatuck: Eno Foundation for Highway Traffic Control.

Scotton, D. W. (1953). "Trends in Rural Retailing in Two Illinois Districts, 1938 to 1950," University of Illinois, *Bulletin*, Vol. 50, No. 71; also Bureau of Business, *Bulletin 76*.

Shakar, Alexious (1958). "A Study of Land Development Along a Major Access Highway With Special Reference to U.S. 99 Between Seattle and Everett, Washington." M.A. Thesis, University of Washington.

Sharpe, Gordon B. (1953). "Travel to Commercial Centers of the Washington Metropolitan Area." Washington: Highway Research Board, *Bulletin 79*.

Smithies, A. (1939). "The Theory of Value Applied to Retail Selling," *Review of Economic Studies*, 6:215-21.

Solow, Anatole A., Allan A. Twichell, and Emil A. Tiboni (1950). *An Appraisal Method for Measuring the Quality of Housing*. Part III: *Appraisal of Neighborhood Environment*. New York: American Public Health Association.

Spokane, City of, City Plan Commission (1952). "Shopping Survey." (Special Report # A.)

------- (1953a). "Population." (City Planning Series # 1.)

------- (1953b). "Employment." (City Planning Series # 2.)

------- (1953c). "Population Distribution." (City Planning Series # 3.)

------- (1953d). "Zoning for Local Business Centers."

------- (1954). "Land Use." (City Planning Series # 4.)

------- (1956a). "Preliminary Draft of Comprehensive Zoning Ordinance. "

------- (1956b). "The Manito Shopping Center Problem. " (Unpublished.)

------- (1957). "Spokane's New Zoning Ordinance. "

Stern, Bernhard J. (1945). *American Medical Practice in the Perspective of a Century*. New York: Commonwealth Fund.

Stewart, John Q. (1950). "Potential of Population and its Relationship to Marketing, " *Theory in Marketing*. Chicago: Cox & Alderman.

Terris, Milton (1956). "Recent Trends in the Distribution of Physicians in Upstate New York, " *American Journal of Public Health*, 46:585-91.

Thaden, John F. (1951). "Distribution of Doctors and Osteopaths in Michigan, " Michigan State College, Agricultural Experiment Station, *Bulletin 370*.

Thompson, Lorin A., and C. H. Madden (1953). "Socio-Economic Relationships of Highway Travel of Residences of a Rural Area. " Washington: Highway Research Board, *Bulletin 67:* 15-21.

Thünen, J. H. von (1926). *Der Isolierte Staat in Beziehung auf Landwirtschaft und Nationalökonomie*. 3rd ed. Berlin: Schumacher-Zarchlin. Translated in part in W. W. Kapp and L. L. Kapp (1949), *Readings in Economics*. New York: Barnes and Noble.

Tinbergen, J. (1957). "The Appraisal of Road Construction: Two Calculation Schemes, " *Review of Economics and Statistics*, 34:241-49.

Traffic Audit Bureau (1947). *Methods for the Evaluation of Outdoor Advertising*. New York: Traffic Audit Brueau.

------- (1950). *Coverage, Repetition and Impact Provided by Poster Showings*. New York: Traffic Audit Bureau.

Troxel, Emery (1955). *Economics of Transport*. New York: Rinehart.

Turvey, R. (1957). *The Economics of Real Property*. London: George Allen and Unwin.

Ullman, Edward L. (1941). "A Theory of Location for Cities, " *American Journal of Sociology*, 46:853-64.

U.S. Bureau of the Census (1952a). *U.S. Census of Housing: 1950*. Vol. 5: Block Statistics, Parts 29 (Cedar Rapids, Iowa) and 183 (Spokane, Washington). Washington: Government Printing Office.

------- (1952b). *Census of Population: 1950*. Washington: Government Printing Office.

------- (1953). *County and City Data Book, 1952*. Washington: Government Printing Office.

U.S. Bureau of Foreign and Domestic Commerce (1924). *Retail Store Location*. Prepared under the direction of L. A. Hansen. Domestic Commerce Trade Information Bulletin No. 269. Washington: Government Printing Office.

------- (1933). *Location Structure of Retail Trade*. Domestic Commerce Series No. 80. Washington: Government Printing Office.

------- (1937). *Intra City Business Census Statistics for Philadelphia, Pennsylvania*. Prepared under the direction of M. J. Proudfoot. Washington: Government Printing Office.

Urban Land Institute, Community Builders' Council (1948). *Community Builders' Handbook*. Washington: Urban Land Institute.

Vidale, Marcello L. (1956). "A Graphical Solution of the Transportation Problem, " *Operations Research*, 4:193-203.

Virginia, University of, Bureau of Population and Economic Research (1951). *The Impact of a New Manufacturing Plant Upon the Socio-Economic Characteristics and Travel Habits of the People in Charlotte County, Virginia.* (Preliminary copy.)

------- (1952). *An Analysis of Changes in the Seasonal Patterns of Road Use from 1949 to 1951 in Charlotte County, Virginia.* (Preliminary copy.)

Voorhees, Alan M. (1958). "Forecasting Peak Hours of Travel. " Paper read before the 37th Annual Meeting of the Highway Research Board, Washington.

-------, G. B. Sharpe, and J. T. Stegmaier (1955). "Shopping Habits and Travel Patterns. " Washington: Urban Land Institute, *Technical Bulletin No. 24.*

Weimar, A. M. , and H. Hoyt (1939). *Principles of Urban Real Estate.* New York: Ronald Press.

Weinfeld, William (1951). "Income of Physicians, 1929-1949, " *Survey of Current Business*, 31:9-26.

Weiskotten, H. G. (1952). "Factors Relating to the Distribution of Physicians, " American Medical Association, *Journal*, 148: 1397-1400.

-------, and M. L. Altenderfer (1940 and 1952). "Trends in Medical Practice-- Analysis of the Distribution and Characteristics of Medical College Graduates, " *Journal of Medical Education,* Vols. 15 and 27.

Weiss, Shirley F. (1957). *The Central Business District in Transition.* Research Paper No. 1, City and Regional Planning Studies. Chapel Hill: University of North Carolina.

Wendt, P. F. (1957). "Theory of Urban Land Values, " *Land Economics,* 33:228-40.

Wheeler, B. O. (1956). *Effect of Freeway Access Upon Suburban Real Property Values.* Seattle: Washington State Council for Highway Research.

William-Olsson, W. (1940). "Stockholm: Its Structure and Development, " *Geographical Review,* 30:420-38.

Winnick, Louis, and Ned Shilling (1957). *American Housing and Its Use: The Demand for Shelter Space.* New York: John Wiley & Sons.

Wynn, F. Houston (1955). "Urban Origin-Destination Characteristics. " University of California, Division of Transportation Engineering. (Mimeographed.)

------- (1956). "Intracity Traffic Movements. " Washington: Highway Research Board, *Bulletin 119.*

Zimmerman, C. C. (1938). *The Changing Community.* New York: Harper.

Zwick, Charles (1957). "Demographic Variation: Its Impact on Consumer Behavior, " *The Review of Economics and Statistics,* 39:451-56.

Index

Accessibility, 145-46, 149. *See also* Travel
Alderson and Sessions, 164
Automobiles: increase in number, 232, 237

Baltimore, 40, 166, 246
Bartholomew, H., 200
Baumol, W. J., and E. A. Ide, 59, 194, 218, 225-26
Beckman, M., 29, 52-53, 234
Beckmann, M., C. B. McGuire, and C. B. Winsten, 163
Benefits of highway improvements, 18-35, 258-60, 264-65, 269-71, 274-75; identification, 19-23, 258-60; vehicular, 20-23, 34, 260; nonvehicular, 20-23, 34, 260, 264-65, 270, 275; transferred, 21-23; reorganization, 21-23, 259, 265, 270, 275; absurd, 25-26; double counting, 25; taxation, 26, 265; double standard, 27-28; negative effects, 28, 260, 270, 275
Breese, G. W., 193
Brink, E. S., and J. S. de Cani, 56, 233
Brownlee, O. H., and W. W. Heller, 23
Burch, J. S., 12-13
Burgess, E. W., 144
Burros, R. H., 53
Business districts: outlying, 40, 49, 232, 249-56; isolated, 40, 41, 44, 49; neighborhood, 40, 41, 43, 46, 47, 49, 252; string street, 40, 42, 43, 48, 49, 57-58, 252; central, 40, 41, 43, 49, 247, 252-56; principal business thoroughfares, 40, 41, 49; community, 43, 48, 49; suburban, 48, 49; regional, 48, 49; nucleated, 58, 75, 77-78, 81, 98-99, 135-38, 208; arterial, 75, 78-79, 81, 99, 118-19, 127-35, 208; auto row, 81, 99, 207; planned, 92-93; centralization tendencies, 118, 136-37, 232; planning for location of, 119-27; empirical definition, 206-7. *See also* Central business district studies

California Division of Highways, 14, 101
Camarillo, Calif., 14-15
Canoyer, H. G., 43-44
Cedar Rapids, Iowa, 68, 86-88, 138, 152-55, 167-76, 198-226, 236, 249
Central business district studies: retail sales attraction, 192; customer buying habits, 193-94; land values, 193. *See also* Business districts
Central place theory: range of a good, 7, 54, 191, 218, 229; Christaller's statement, 50-

51, 190; hexagonal market areas, 51, 54; Beckmann's city size model, 52-53; control ratio, 53; Burros' theory of static hierarchy, 53; threshold, 54, 229; and theory of tertiary activity, 54-55; hierarchical marginal good, 55; hierarchy, 149, 190, 218. *See also* Location theory
Centralization of retail business, 118, 136-37, 232
Charlotte County, Va., 166
Chicago, 40, 44, 246
Christaller, W., 50-51, 190
Cincinnati, 68, 88-92, 93-94, 138, 246
Ciocco, A., and I. Altman, 241-43, 247-49
Concentric zone theory, 144
Consumer behavior: retail shopping model, 59, 164, 194, 218, 225-26; and demographic variations, 165. *See also* Travel
Customer movements, 209-10; households to business establishments, 212-19; business establishment to business establishment, 219-23. *See also* Travel; Movement of individuals

Demand for travel or transportation, 6, 16-17, 23, 26-27, 30-32, 158-63; as a function of traffic conditions, 163; models of, 170-76; by business type visited, 222
Denver, 49
Detroit, 41, 246
Dickinson, Frank, 235, 237-38

Elapsed time: in relation to trips away from home, 161; model of, 173
Enke, S., 233
Everett, Wash., 94, 101, 103, 109, 138

Fort Wayne, Ind., 167
Fountain Square, Evanston, Ill., 44
Friedmann, J. R. P., 16

Galpin, C. J., 12
Gardner, J. C., 166
Garrison, W. L., 44-45, 148, 156; and M. E. Marts, 14, 101
Grebler, L., E. M. Blank, and L. Winnick, 142-43

Haig, R. M., 185
Hamburg, J. R., 166
Harris, C. D., and E. L. Ullman, 145
Highway capacity: and resource utilization, 5,

HIGHWAY ECONOMIC STUDIES

Available from Department of Civil Engineering, University of Washington:

State Interest in Highways: A Report on Highway Classification. Vols. I and II. Staff, Washington State Council for Highway Research, R. G. Hennes, Chairman, Willa Mylroie, Executive Secretary. 1952.

Nature of Highway Benefits: A Progress Report on a Study of Tax Responsibility for Meeting the Costs of Public Roads. Staff, Washington State Council for Highway Research. 1952.

Feasibility of Toll Roads in Washington. Staff, Washington State Council for Highway Research. 1952.

Natural Resource Roads. Staff, Washington State Council for Highway Research. 1952.

First Biennial Report. Staff, Washington State Council for Highway Research. 1952.

Second Biennial Report. Staff, Washington State Council for Highway Research. 1954.

Allocation of Road and Street Costs. 1956.

Part I. *An Equitable Solution to the Problem.* R. G. Hennes *et al.*

Part II. *Classification, Traffic Volumes, and Annual Costs of County Roads and City Streets.* G. A. Riedesel, R. S. Turner, and E. B. Slebodnick.

Part III. *Bases for Weight-Distance Taxation in the State of Washington.* J. W. McGuire.

Part IV. *The Benefits of Rural Roads to Rural Property.* W. L. Garrison.

Part V.. *The Effect of Freeway Access upon Suburban Real Property Values.* B. O. Wheeler.

Part VI. *Commercial Motor Carriers as Highway Users in Washington.* S. H. Brewer, R. B. Ulvestad, and P. T. McElhiney.

Vol. I. *Competitive and Structural Analysis of Commercial and Highway User Industry in Washington.* Vol. II. *The Volume and Character of Freight Traffic Movements by Commercial Motor Carriers in Washington.* Vol. III. *Factors Relating to Earning Values of Commercial Motor Carriers in Washington.*

Exemptions and Diversions from Motor Vehicle Licensing and Taxation. M. Ekse and R. H. Myers. 1957.

Monthly Licensing for Seasonally Operated Motor Vehicles in Washington. W. I. Little and S. H. Archer. 1958.

Influence of Highway Improvements on Urban Land: A Graphic Summary. W. L. Garrison and M. E. Marts. 1958.

State Pre-emption of Highway User Taxes. L. D. Goldberg and R. W. Lambright. 1958.

Geographic Impact of Highway Improvement. W. L. Garrison and M. E. Marts. 1958.

Equitable Alternatives in Highway User Taxation. R. G. Hennes and Richard L. Pollack. 1958.

Determination of Special Benefits Resulting from Highway Location. Warren R. Seyfried. 1958.

Urban Freeways: Their Development and Financing in Washington State. E. M. Horwood, L. D. Goldberg, and Donald F. Rieg. 1959.

Available from the University of Washington Press:

Studies of Highway Development and Geographic Change. W. L. Garrison *et al.*
  1959.
Studies of the Central Business District and Urban Highway Development. E. M.
  Horwood *et al.* (In press.)